DISCARD

Risk Management in Agriculture:

A Guide to Futures, Options, and Swaps

Join us on the web at

agriculture.delmar.com

Risk Management in Agriculture:

A Guide to Futures, Options, and Swaps

Lowell B. Catlett

Dean, College of Agriculture and Home Economics
Regents Professor of Agricultural Economics
New Mexico State University

James D. Libbin

Professor, Department of Agricultural Economics
and Agricultural Business
College of Agriculture and Home Economics
New Mexico State University

THOMSON

DELMAR LEARNING

Australia Canada Mexico Singapore Spain United Kingdom United States

THOMSON

DELMAR LEARNING

Risk Management in Agriculture: A Guide to Futures, Options, and Swaps
Lowell B. Catlett and James D. Libbin

Vice President, Career Education Strategic Business Unit:
Dawn Gerrain

Director of Learning Solutions:
Sherry Dickinson

Managing Editor:
Robert L. Serenka, Jr.

Acquisitions Editor:
David Rosenbaum

Product Manager:
Gerald O'Malley

Editorial Assistant:
Christina Gifford

Director of Production:
Wendy A. Troeger

Production Manager:
JP Henkel

Senior Content Project Manager:
Kathryn B. Kucharek

Director of Marketing:
Wendy Mapstone

Channel Manager:
Gerard McAvey

Cover Image:
Corbis

Cover Design:
Kristina Almquist

For permission to use material from this text or product, submit a request online at http://www.thomsonrights.com Any additional questions about permissions can be submitted by email to thomsonrights@thomson.com

Library of Congress Cataloging-in-Publication Data

Catlett, Lowell B.
  Risk management in agriculture: a guide to futures, options, and swaps / Lowell B. Catlett, James D. Libbin.
      p. cm.
    Includes index.
    ISBN-13: 978-1-4018-1441-0
    ISBN-10: 1-4018-1441-7
    1. Futures–United States. 2. Agricultural prices–United States. 3. Risk management–United States. 4. Options (Finance)–United States. 5. Commodity futures–United States.   I. Libbin, James D. II. Title.
      HG6024.U6C38 2006
      630.68'1—dc22

2006013234

NOTICE TO THE READER

Publisher does not warrant or guarantee any of the products described herein or perform any independent analysis in connection with any of the product information contained herein. Publisher does not assume, and expressly disclaims, any obligation to obtain and include information other than that provided to it by the manufacturer.

The reader is expressly warned to consider and adopt all safety precautions that might be indicated by the activities herein and to avoid all potential hazards. By following the instructions contained herein, the reader willingly assumes all risks in connection with such instructions.

The Publisher makes no representation or warranties of any kind, including but not limited to, the warranties of fitness for particular purpose or merchantability, nor are any such representations implied with respect to the material set forth herein, and the publisher takes no responsibility with respect to such material. The Publisher shall not be liable for any special, consequential, or exemplary damages resulting, in whole or part, from the reader's use of, or reliance upon, this material.

For Joni and Gail

Contents

Preface

Agricultural businesses have used traditional futures contracts since post–Civil War times to help manage the risk of price changes. Throughout the last century, various new futures contracts were added and several discontinued as the dynamics of the agricultural industry changed. However, from the late 1930s until the early 1980s futures contracts were the only form of derivatives available as options (called privileges) were banned by the 1936 Commodity Exchange Act. Options on futures contracts were reborn in the early 1980s and have continued to grow in popularity ever since. Swap derivatives did not emerge until 1979 but quickly gained market support.

Risk management today, more so than at any time in history, has an abundance of different tools that can be to used to mitigate the risk of price change. This abundance of possible choices requires that modern risk managers must not only be familiar with all of the choices, but understand how they might be used in a comprehensive risk management program.

The book approaches the use of the various derivatives from an overall risk management viewpoint versus the more traditional single event course of action. Traditional futures use by farmers, for example, concentrated on forecasting where prices might move and then selecting a futures contract to hedge the risk of the price movement. Yet the vast amount of price forecasting research during the last few decades shows that markets are efficient and thus almost impossible to forecast. Furthermore, the focus of business management during the same time points to the importance of valuing the risks that businesses face and determining how or if the risks can and should be managed.

Today risk managers have to decide not only which tools to use—futures, options, swaps, or an infinite combination of the different tools—but when and if the tools should be used as part of the overall risk of the business activity. The derivatives markets are robust and teeming with speculators willing and able to accept the risks that agricultural businesses face. The responsibility of the modern risk manager is to supervise the use of the speculators via various derivatives to enhance overall profitability. This book is an attempt to help risk managers achieve that goal.

The authors wish to thank Irma Marshall for her outstanding skills in getting this book into a readable format. A hearty thank you to the kind people at Thomson Delmar Learning who maintained their deep well of patience and professionalism despite our numerous attempts to drain it dry. A special thanks goes to the various reviewers who helped keep us on track and made every effort to correct our errors. The errors that remain are ours alone.

Introduction

KEY TERMS

weather risk	margin	option
biological risk	margin call	premium
production risk	offset	strike price
marketing risk	roundturn	two-step derivative
policy risk	future	swap
strong contract	derivative	counterparty
weak contract	one-step derivative	notional
standardized contract		

This chapter gives a general overview of agricultural risks and an introduction to the basic tools to manage market risk—futures, options, and swaps.

OVERVIEW

Agriculture challenges the body and soul. A farmer battles Mother Nature and gets a crop produced only to find the market isn't offering enough to cover his cost of production. A rancher gets his cows through a difficult calving season because of a cold wave, only to face a mid-summer drought with little grass for the calves. The calves are weaned early, and the market price is so low that the rancher cannot cover his supplemental feed costs, much less make his land payment. A milk cooperative bottles whole milk only to find that customers shun the product because of a recent television show about health concerns. The price of highly perishable milk has to be severely discounted to move it out of the stores. An agricultural equipment manufacturer finally completes a major sale to a foreign country after months of negotiations and help from export groups and the federal government. The manufacturer is paid in the foreign currency. By the time the currency is exchanged, the exchange rate is unfavorable and despite the best efforts of lots of people and an incredible amount of time and money spent on the effort, the foreign exchange loss makes the deal unprofitable.

And yet, agriculture invigorates as well. A farmer plants a new short-growing-season corn variety that ultimately is harvested early enough to get a 10 cent per bushel premium above the normal harvest time price. A rancher, noting the position of the cattle cycle, sells her feeder cattle earlier than normal and at a lighter weight, but receives a higher total price for each animal than if she had waited to sell at the normal time. A cheese producer shifts 10 percent of the plant capacity to a new slower melting variety with the anticipation of signing a new contract with a large frozen pizza manufacturer with a 15 percent price premium. A farm equipment manufacturer reduces production of a major equipment line and increases the production of a new piece of equipment used for minimum till operations to be in position for increased sales when changes in farm legislation take effect.

Risks are everywhere and they can be both positive and negative. Risk management isn't just about reducing or mitigating potential problems, it is also about choosing, evaluating, and accepting the right kinds of risks so that a profit can be obtained and maximized.

Agricultural Risks

Early frosts and droughts are hazards that farmers and ranchers know only too well. **Weather risk,** despite more than 100 years of trying to overcome it with technology, can hurt producers' incomes. By the same token, weather can provide agricultural producers with a bounty. Rain at the right time is a blessing, but during harvests a curse. An old French saying puts it best: "The rain that waters the field muddies the road." A rain on the fourth of July can wreck a parade and a downed hay crop, but save a drought-stricken rancher. **Biological risks** include a major pest that destroys all or part of a crop and bacteria that invades a cheese manufacturing process only to ruin a large vat of product. However, the right kinds of enzymes are necessary for cheese production, and certain insects or microbes are necessary and beneficial. Botrytis mold causes grapes to rot, yet in the Bordeaux region of France, when wine makers learned that the mold causes the sugars to concentrate, one of the great dessert wines of the world was born—Sauterne. When botrytis is managed properly, great things emerge; when it isn't, the crop is lost.

Weather and biological risks are the two most important natural risks that agribusinesses face. These two natural risks are generally included in a more universally used term for farmers and ranchers called **production risk.** However, farmers and ranchers are also confronted with numerous other risk forms such as marketing and policy risks. **Marketing risk** includes not only movements in prices, but also changes in all of the other components of the market channel such as contracting terms, grades and standards, and sometimes the very simple issue of having an opportunity to sell the product. Typically marketing is thought of as finding the highest price, yet the channel elements are every bit as important. The first handler of a raw agricultural product, such as a grain elevator, has to procure the grain via cash spot buying or some form of forward contracts. How they end up buying the grain will determine some of the risks they will face when they sell into the next market channel. The elevator will also have to decide what grades of grain to sell and if they should "blend" for certain grades, and what storage risks such as moisture and pest problems need to be handled. By the same token, a rancher also faces the decision regarding at what weight her animals need to be sold and how the animals will quality grade.

During the late 1970s the state of Alaska decided to develop a large block of land southeast of Fairbanks in the Delta Junction region for agricultural production. The climate was suitable for small cereal grains, mainly barley, and grasses for livestock. The land was sold to farmers and cleared and put into mostly barley production. The state promised assistance in developing an infrastructure to market the barley including rail lines and port facilities, because there was no place to sell the crop once produced. Budget shortfalls by the state stopped any development of a market infrastructure. Farmers now had cleared the land, developed the property as farms, and produced crops that could not be sold because there were no market channel participants to be the first handlers of the product. The Delta Junction farmers had considerable market risk as well as **policy risk.** The state of Alaska developed an agricultural policy but only completed a portion of the project. Some policy risks are political and/or social and therefore not directly commodity specific, such as a policy that protects a particular classification of land, like watersheds. Certainly some policy risks are economic in nature. The state of Alaska's problem with the Delta Junction agricultural project resulted from a major decline in energy prices in the early 1980s that slashed state revenues. The state, despite its promise to develop a market infrastructure, did not have enough money to complete the project. Unfortunately, the Delta Junction farmers bore the entire policy risk, which destroyed most of the farming operations. The flip side would have been a boom for the Delta Junction farmers. A marketing infrastructure would have created a place to sell their crops and their farms would have increased in value relative to the surrounding, undeveloped scrub timberland.

The risks that agriculture confronts are, of course, too numerous to fully analyze here. For the purposes of this book marketing price risks will be the major focus, yet the other major risks involving weather, biology, and policy will come to bear on many of the marketing risks. No decision is ever totally a marketing event nor exclusively financial or managerial. Every business decision will have components of all business functions and likewise no marketing risk is totally devoid of other risks. The major goal of this book is to frame marketing price risks within the larger framework of other agricultural risks so that better and more informed decisions are possible for managers.

Managing Agricultural Risks

Risks are managed by either proactively entering into the process of influencing the outcomes or accepting the outcomes without interference. Accepting the outcomes is not the same thing as ignoring them. To ignore a risk is to be ignorant of its existence or its consequences. To accept a risk is to acknowledge the potential outcomes and be willing to accede to their impacts. Accepting a risk without trying to influence the outcomes may arise from knowing that you as an individual cannot change the outcome (weather), the cost would be unacceptable, at least relative to the potential losses (the price of insurance), or the probability of the event happening is perceived as low and therefore unlikely to occur (a comet crashing into the business). Management certainly involves the acceptance of risks and without question also includes ignoring risks as not every risk can be anticipated or analyzed. Proactively managing risks is the process of looking at the probability of the event occurring, what the potential outcome might be, and how that outcome might change if certain risk management tools were used. Oftentimes this process of risk management is called, quite simply, management.

Production Risks

Consider a wheat producer who knows that in one out of every five years he won't get enough rainfall to make a crop. Irrigation may be possible, but the equipment must be purchased and the producer must learn how to properly irrigate the crop. The producer proactively manages the weather risk of not having adequate rainfall by installing irrigation equipment. Another producer in a similar position may decide to accept the risk because she may view the added expenses of irrigation as not economically justified or she doesn't want to learn the process of irrigation for such a small probability of occurrence. Or she may decide to proactively manage the risk, not by installing irrigation, but by switching to a more drought-tolerant crop or wheat variety. The drought-tolerant variety might yield a lower production level in normal years, but a higher amount during droughts. In either case, each producer has considered the probability of the event occurring and what the outcomes might be, and has decided to manage those potential outcomes in different ways. Still another producer might opt for purchasing crop insurance as a method of production risk management.

Pests could be managed on a farm or ranch via a preventive insecticide spraying program or the release of beneficial insects to destroy invading species. A food manufacturer knows that washing a piece of equipment with a sanitizing solution every third use will kill

99.99 percent of harmful bacteria and that process meets regulation requirements. They also know that washing after every use has a 100 percent kill factor. Is the difference in kill factors for harmful bacteria worth the added expense? Good risk management attempts to quantify potential losses and determine if the risk is worth taking.

Weather and biological risks can in many cases be proactively managed as the previous examples point out. Similarly, many production risks have to be accepted as the cost of being in a particular business. A farmer might have supplemental irrigation for droughts, tree bands for wind erosion, and a mix of varieties for added drought protection only to have a tornado destroy the farm. Food producers who run 24-hour production lines are prone to brownouts or blackouts for electrical power, as several had to deal with in 2001 in California. The processors thought they had on-site protection in the form of supplemental electrical generators, but were informed by the state's air quality regulators that they could not use the standby generators without a substantial daily fine. Not all risks can be proactively managed and merely have to be accepted.

Policy Risks

Farmers found out the harsh realities of world trade when President Carter imposed a ban on grain shipments to the former Soviet Union in 1980. Grain prices tumbled. Such policy risk is virtually impossible to manage. In 1972 when secret deals to sell grain to the former Soviet Union were revealed, grain prices soared, but most producers didn't benefit because the information was not generally known in the market place. A new policy was put into effect that required that certain sized grain deals had to be made public within a very short period of time so that farmers could reap some of the benefits of rising prices. Secret or unknowable decisions that impact policy cannot be effectively managed. However, many other policy risks lend themselves to being proactively managed such as support price levels for certain grains, export enhancement programs for agribusinesses, and leasing rates for public grazing lands. Public policies that affect food production, processing, and marketing are subject to change, but the process of change is done in the public arena, thus information is available on what the potential changes might entail.

Marketing Risks

Marketing risks are composed of two major risks: structural risks and price risks. Structural risks are those risks associated with the market channel such as how many sellers/buyers exist, what is the competitive level, and what is the accepted way to package, weigh, transport, grade, and regulate. Price risk is simply the risk of price movement.

Structural Risks

Farmers who do not have enough (or any) choices to sell their products have formed cooperatives to act as the first handler of the raw agricultural product and for certain processing steps when the competitive structure didn't offer many choices. Grain and dairy cooperatives exist throughout the country because farmers wanted to have more control over the market channel for their products. Livestock producers started selling their cattle via electronic means, first with teletypes and later with computers and videos, as a way to bypass a local auction house or only a few order buyers. Food processors have vertically integrated into transportation and wholesaling as ways to assure that their product quality is maintained.

Alice Waters has numerous farmers supplying fresh products for her restaurant Chez Panisse in Berkeley, California, so that she can assure her patrons the very freshest and best produce available. Ms. Waters manages the risk of getting the freshest and best products for her restaurant via direct contracts with farmers, bypassing the traditional restaurant supply channel. Pacific Foods, Portland, Oregon, contracts directly with soybean producers to avoid the risk of not getting the quality they need for their food production company.

Price Risks

The major difference between all of the other major agricultural risks previously discussed and price risk is that all of the other risks are event-participant specific. How a farmer overcomes the lack of enough buyers for his strawberries will be different from how a rancher deals with a change in grazing fees on public lands. Each risk is unique to the participant and the particular risk. A change in agricultural policy may cause an increase in the price of corn, but there is no general tool available to combat the risk of the policy being put into place. Irrigation is a general tool to counter the effects of drought risk, but it cannot be used by every producer who faces the risk nor is it always available. Price risk is universal in agriculture and therefore general tools to control the risk of price movement exist.

The risk of receiving a price that does not cover the cost of production, processing, or transporting a product is a major problem within agriculture. Furthermore, the risk of paying more for an input than is justified economically is a significant risk. Fortunately, several price management tools exist that can be applied to a broad set of price risk management needs.

Tools of the Trade

One major way to begin the process of managing a risk is the process of *securitization*. Securitization is the procedure to identify, quantify, and structure a risk into a financial instrument (security). Market traders then determine the value of such a security. Home mortgages have long been securities with individual default risk but they lacked a broader market until the Government National Mortgage Association (GNMA) was formed to package individual home mortgages into a larger security that was attractive to institutional investors; Ginnie Mae became the security that managed the risk of individual mortgage default. The securitization of individual home mortgages has been a great success story in the last 30 years that has been copied by Fannie Mae and Farmer Mac.

Agricultural production risks have been securitized by the introduction of crop insurance policies and, as of 2004, livestock production insurance policies. Life insurance has long been a process of securitization of the loss of income from the death of an individual by providing a lump sum payment in the event of death.

Often the securitization process is done in a roundabout way. In the glory years of the infamous company Enron, the firm once constructed a security for orange juice processors in Florida. Orange juice processors could not secure long-term contracts for electricity. Enron offered them longer term contracts for the electricity than they were getting, with step adjustments based on the price of orange juice. If juice prices went higher, the processor would pay a higher electric rate when the step adjustment period arrived, and vice versa. Enron then hedged the risk with orange juice futures contracts. Enron allowed the processors to securitize the risk of electrical rate changes.

Weather risk has been securitized during the last few years (since 1997) by using a process known as heating and cooling degree days. A bench mark temperature such as 65 degrees Fahrenheit is used. The amount the daily average temperature deviates from the bench mark is the number of heating degree days (HDD) if under and cooling degree days (CDD) if over. For such a number to have value for an industry a security must be developed that can be correlated with known risks. Obviously, utilities would find such a security valuable as HDDs and CDDs would be strongly correlated with power needs for either heating or air conditioning needs. The Chicago Mercantile Exchange has started offering weather derivatives (2005) with the hope that other industries, such as agriculture, that have weather risks will find the new securities to have value as a weather risk management tool.

Price risk has been securitized for many years in the form of forward, futures, options, and swap contracts because price is easy to identify, quantify, and structure into a security. Each of these securitized tools has a unique way to handle the risk of price change.

Forward Contracts

One of the most used tools for controlling price risk is a simple forward contract. Two parties agree upon contract terms and set a price for the product. Both parties know the price in advance and therefore have no risk that the price will change. Some forward contracts can be traded to another party. If a forward contract has a retrade clause, the contract is said to be **strong** but if it lacks a retrade clause it is said to be **weak.** Residential home mortgages are retraded, often several times during the life of the loan, and are thus strong contracts. On the other hand, the contract between an individual and a home builder is generally a weak contract. The home builder must build a home for the contracting party and cannot sell that obligation to someone else. The contract can be for any product or service and specify details about the size, grade, and any points that both parties agree. Because each contract is unique, forward contracts are nonstandardized.

Forward contracts are in wide use in almost all areas of agricultural production and agribusinesses. A forward contract reduces (but does not necessarily eliminate) the risk of finding a market for a product. If price is part of the contract, then by entering the contract the contracting party that has something to sell has eliminated the risk of price change. By the same token, some risks still exist and in fact may increase. One of the parties involved in the contract my not fulfill the terms of the contract. Therefore, one of the major risks of forward contracts is *default risk*. If the party that wants the product delivered, i.e., the buyer, finds that the seller has reneged (defaulted), it may be very difficult to find another seller quickly or with the quality needed. If market prices move substantially from the contracted price, the business has lost a major opportunity to profit from the price move because they had a forward contract.

Futures Contracts

A futures contract is traded on a central exchange that is (in the United States) regulated by the Commodity Futures Trading Commission (CFTC). Contract terms are prespecified as to size, grade, delivery locations and times, and other important details and are **standardized.** The contract's price is determined by an open outcry auction at the exchange. All futures contracts are retradable (strong). The buyer of the contract has agreed to accept delivery of the item and the seller has agreed to deliver. In the case of items that can't physically be delivered such as a stock index, the buyer and seller simply use a cash price at delivery to cash settle the terms of the contract and no physical delivery occurs. Some items that could be physically delivered such as feeder cattle are cash settled and no actual delivery takes place.

All futures contracts are leveraged contracts. The buyer and seller put up a fraction of the contract's value (**margin**), usually about 10 percent, to gain control of the contract. As the contract changes in value, the traders will have to pay in or are allowed to take out any paper losses or gains. The process of paying paper losses is called making a **margin call.** If paper gains occur, the contract holder can call for the extra margin to be returned within limits. Margin calls are necessary because the trader has only put up a fraction of the value of the contract and as the contract loses value on paper, those losses will have to be paid to assure that the trader has the financial solvency to hold the contract.

Since all futures contracts are standardized, strong, and traded on centralized exchanges via an open system of outcries, the contracts are very *liquid*. The buyer or seller of a contract can have an order filled very quickly, usually within three minutes. They can get out of the contract's obligation by **offsetting.** The buyer of a contract would offer to sell as a way to offset and an initial seller would offer to buy the contract. A buy-sell, sell-buy constitutes a **roundturn.**

Futures contracts are used to manage risk two ways. First, because they are legally binding contracts they can be used as a way to sell or buy a cash commodity. If a food processing company buys a March futures contract on corn, they could wait until March and demand delivery of corn at the contract term's 5,000 bushels of #2 yellow corn and they would pay the price specified in the original contract. If the March futures contract was purchased in November of the previous year, the firm fixed the price the buyer would have to pay for the corn over four months in advance. Even for cash-settled futures contracts the effect is similar to actual delivery—the actual price is fixed, but the product is not delivered. The use of a futures contract to actually buy or sell a cash commodity is usually low (averaging 5 percent of total futures contracts traded, but sometimes reaching 20 percent or more). The second use of futures to manage risk, *hedging*, is the one most often used. A March corn futures is purchased in November by a food processing company. In March, they buy the actual corn in the cash market and offset the original buy in futures market by selling the contract back. The amount of money they made or lost on the round turn is used to add to or subtract from the actual cash price. Hedging allows the business to operate within the normal cash markets and futures markets simultaneously and manage the risk of price change.

A futures contract derives its value from the underlying cash commodity, thus it is said to be a **derivative** contract. Since the value of the futures contract is directly tied to the cash commodity it is said to be a **one-step derivative.**

Options Contracts

The buyer of an **option** has the right but not the obligation to do something. The seller of the option has the obligation to perform as specified in the agreement. If the buyer has the right but not the obligation to buy something, the option is referred to as a *call*. If the buyer has the right but not the obligation to sell something, the option is a *put*. One way to remember puts and calls is as follows: when you sell you are putting something away from you or you put something on to the market; and when you buy, you are calling something in to you or you call something from the market. For the right but not the obligation to buy or sell something, the buyer pays a **premium.** The price at which the product or service is exchanged at some point in the future is called the **strike price.** Two types of options exist: (1) exchange-traded options on futures contracts, and (2) options on physical commodities (the actuals), also called OTC (over the counter). Options on futures are traded at each of the futures exchanges and are standardized just like the underlying futures contracts. Options on the physicals are nonstandardized, just like typical forward contracts. One of the more popular forms of options on the physicals, a real estate option contract, is used extensively in real estate sales. A potential buyer of a piece of property negotiates with the owner to give him an exclusive opportunity or option to buy the property within a certain period of time at a certain price (strike price) and the potential buyer pays the current owner (seller) a premium. The contract involves an exclusive right to buy—the seller cannot sell the property to anyone else with the specified time period and receives the premium as the consideration for allowing the buyer to decide if he really wants the property (including whether the buyer can put together a financing package to purchase the property). The buyer has the right but not the obligation to buy the property within the specified time period at the strike price. If the buyer decides to buy the property, he *exercises the option* and buys the property at the strike price. Usually in real estate deals the option premium is then deducted from the strike price, but not always. It is a point of negotiation. If the buyer decides not to exercise the option, he *lets the option expire* and doesn't have to buy the property; the seller is free to sell to someone else and keeps the premium. An option on a futures contract is mechanically the same as an option on the physicals. The only major difference is that the item that is bought or sold is a standardized futures contract. The value of the underlying futures contract determines the value of the options contract, thus an option on a futures contract is double derivative, or a **two-step derivative,** because the futures contract's value was derived from the cash commodity's value and the value of the options contract is derived from the futures contract.

Swap Contracts

A **swap** entails the exchange between two or more parties (called **counterparties**) of the cash flows arising from other contracts or entities (called **notionals**). The individuals who put swaps together are called *swap dealers* or *brokers*. A broker simply lines up the counterparties and receives a fee for services rendered. A dealer usually constructs one deal with one counterparty and then finds another counterparty to construct another deal. The dealer is rewarded by the difference in the deal terms. Most swaps are put together by dealers rather than by brokers because dealers can book one side of the deal immediately whereas a broker has to find two counterparties simultaneously to get the deal booked. However, the dealer faces the risk of finding another counterparty after they have booked one counterparty.

For an example of a swap, consider a food manufacturer who is currently buying cash corn on a weekly basis in the cash market. Some weeks he pays more than the previous weeks, some weeks he pays less. Each week he faces the risk of the price of corn increasing. If the food manufacturer entered into a swap with a swap dealer, the deal might be to continue buying the corn in the cash market. Although the food manufacturer would continue paying the market price each week, he would pay the dealer a fixed price set in advance not subject to change for the next six months. The dealer in turn would send a payment to the manufacturer each week equal to the current cash price and in return he would receive a fixed price payment from the manufacturer. The manufacturer and dealer have swapped cash flows. The manufacturer has swapped paying a variable price each week for a fixed price each week for the next six months. Consequently, the manufacturer has eliminated the risk of having to pay a different price each week, exchanging that uncertainty for the certainty of paying a fixed price. Recognize that the manufacturer still obtains the corn each week and pays the market price each week for the corn; however, because he receives the market price each week from the swap dealer, the cash market deal is a total wash. The manufacturer ends up paying to the dealer a fixed price each week for the next six months.

The dealer has now absorbed or accepted the weekly risk of price changes in the corn market. The dealer has to pay a variable price each week to the manufacturer but receives a fixed price. For the dealer to manage the price risk, he will need to find another counterparty that wants just the opposite. The dealer could offer to a grain elevator the following deal: sell your corn each week in the cash market and receive the cash price. Pass the weekly cash price on to the dealer and in return the dealer will pay a fixed price each week to the elevator for the next six months. The elevator sells its corn each week into its normal market channels and receives the weekly cash price which it passes to the dealer. The dealer then sends the elevator a fixed price each week. The elevator has eliminated the risk of weekly price change. The dealer will make a profit based on the difference between the fixed prices received from the manufacturer and the elevator. The weekly cash price is merely a pass through and a total wash. This type of swap is called a *commodity swap*. Two more types of swaps are used by various businesses—*interest rate swaps* and *exchange rate swaps*. Interest rate swaps exchange a fixed rate for a floating rate or vice versa between counterparties that have bank loans. Exchange rate swaps offset exchange rates by allowing counterparties to move from one currency to another and do so at a known exchange rate.

Swaps allow for the shifting of price risk via cash flow exchanges. The counterparties do not swap the underlying asset, only the cash flows that the assets or business activities generate. They differ from futures and options in that they are not standardized contracts and are not exchange traded. Table 1–1 compares the price risk management tools' various characteristics.

Using the Tools

Hedging

If futures, options, and swaps are used to manage price risk the process is called hedging. The term hedging is loosely applied to many operations that try to control risks. The process of hedging includes having some market situation that bears a business or economic risk and then trying to offset or control that risk by having a position in another market that would mitigate the risk, should it occur, in the cash market.

Table 1–1   Price Risk Management Tool Comparison

	Forwards	Futures	Options on Futures	Swaps
Exchange Traded	No	Yes	Yes	No
Standardized	No	Yes	Yes	No
Default Risk	Yes	No	No	Yes
Leveraged	No	Yes	Yes	No
Liquid	No	Yes	Yes	No
Price Determination	Individual Negotiation	Open Competitive Process	Open Competitive Process	Individual Negotiation

Education and on-the-job work experience are forms of hedging. The market place does not reward unskilled workers with bountiful wages, but it does reward successful entrepreneurs, successful athletes, and successful TV or movie stars, regardless of education level. Education either through a college degree, apprenticeship, or training is a way to hedge against a low wage rate market. Skilled and educated workers earn higher wages than unskilled workers or unsuccessful entrepreneurs, athletes, or actors. A farmer who produces both crops and livestock has hedged. If a crop failure occurs, then hopefully the livestock operation will help offset the loss. Farmers who diversify crops are hedging. The loss of either yield or price in one crop might be offset by a gain or yield or price in another.

Hedging is simply having two or more positions in different markets so that the loss in one is offset with a gain in another. Chapters 4, 5, and 6 cover each of the several hedging tools in detail, but a simple hedging example using the most popular price risk management tool—futures contracts—will be used to illustrate the fundamental concept.

Corn Production Hedge Example

A corn producer plants her crop in the spring and harvests during the mid fall. Once the producer has started growing corn, she has a financial risk. If the price she receives in the fall for the product is not high enough to cover her cost of production, then she will suffer a loss. In the springtime if she estimates that her cost of production will be $2.00 per bushel and she observes that the local cash price is currently $2.25 per bushel she would have a profit margin of $0.25 per bushel—if she had corn to sell in the spring. Her risk is that a price of $2.25 per bushel in the spring will not be the price in the fall when she will have a crop to sell. She would of course be very happy if the price was above $2.25 per bushel, but the downside risk is that it will be lower than $2.25 per bushel and her expected profit margin squeezed. If the price drops below $2.00 she will have a loss.

The farmer can do nothing to control the movement in the corn market price. What the producer needs is some financial asset that would gain in value when the corn price moves down. If a December corn futures contract was sold in the springtime when the crop was planted at a price of $2.35 per bushel and the price decreased, then the contract could be offset by buying it back at a lower price with a profit on the futures trade. This profit could be used to offset the loss in the cash market. Figure 1–1 shows the effects of a simple production hedge.

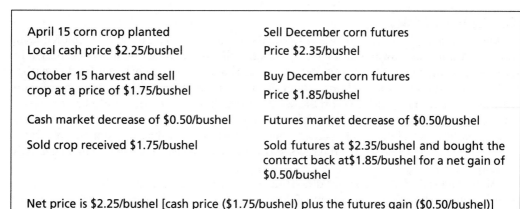

April 15 corn crop planted Local cash price $2.25/bushel	Sell December corn futures Price $2.35/bushel
October 15 harvest and sell crop at a price of $1.75/bushel	Buy December corn futures Price $1.85/bushel
Cash market decrease of $0.50/bushel	Futures market decrease of $0.50/bushel
Sold crop received $1.75/bushel	Sold futures at $2.35/bushel and bought the contract back at$1.85/bushel for a net gain of $0.50/bushel
Net price is $2.25/bushel [cash price ($1.75/bushel) plus the futures gain ($0.50/bushel)]	

Figure 1–1   Corn production hedge

The hedge protected the corn producer from the declining cash market price. The producer received a net price equal to the cash price at the time of planting the crop.

This hedge example shows the cash and futures markets moving together. When one decreases the other does likewise. The economic reason that this occurs is discussed fully in Chapter 3. Suffice to say at this point that the cash and futures markets *tend to trend together* and that trend likelihood allows the cash market to be managed with a futures contract.

Speculation and Income Generation

Futures, options, and swaps can be used to simply speculate on market price direction or price relationships between two or more markets. Traders who want to guess (speculate) on which direction prices will move can also use futures and options contracts in addition to hedgers. The example in Figure 1–1 shows a hedging example, but if the cash portion of the example were removed, the futures transactions would be a simple speculation that the price of corn was going to decline. In the example the price declined and the futures trades yielded a gain of $0.50/bushel. If the price had gone up instead, the futures transactions would have resulted in a loss. Recognize also that if the corn producer had not hedged, then the cash portion of the example is speculation. The corn producer is speculating when she plants her corn that the price will be high enough at harvest to cover her costs and make a profit. Traders or businesses who try to guess in which direction price will move are called *position* or *directional speculators*. Recognize that any business activity that has a price risk that is not managed via hedging is *de facto* speculating with the cash price movement.

How does the price of corn in Iowa relate to the price of corn in Texas? Is there a relationship between the value of soybeans and the value of cottonseed? Traders who attempt to profit from knowledge about the relationships between two or more markets are generally known as *arbitragers*. Futures, options, and swaps can all be used either by themselves or in combinations to profit from arbitrage opportunities.

Position and arbitrage speculation will be discussed at length in various chapters as they pertain to certain risk management concepts and strategies. Knowledge of speculation is critical to successful hedging and risk management, if for no other reason than to simply know when a financial risk is being accepted, i.e., speculation, or when it is being managed—hedging.

The Role of Risk

Certain risks in agriculture can be managed. Marketing risks can be directly handled using futures, options, and swap contracts. Other agricultural risks can be managed only indirectly, but surprisingly, futures, options, and swaps have a role to play in managing weather, biological, and policy risks. Major droughts can be a factor in causing the price of commodities to increase, just as favorable weather conditions can contribute to decreasing prices. Consequently, major weather risks can show up indirectly as price risks. The same goes for policy risks, biological risks, and other risks that individuals cannot control, yet as individuals the consequences of the risks show up as price risks that can be managed individually via hedging.

Managing risk isn't just about offsetting a potential loss, it is also about producing a potential gain. The role of speculative risk to generate income is also extremely important to agriculture and risk management in general. The remaining chapters will cover in detail how to use the tools of futures, options, and swaps to manage agricultural risk—both for loss protection and income generation.

CHAPTER 1—QUESTIONS

1. What are some major differences between forward contracts and futures contracts?

2. What are some of the major differences between swaps and futures contracts?

3. What is the reason for margin requirements on futures contracts? Should hedgers be concerned about margin calls? Why?

4. Why are futures and option contracts called derivatives?

5. What are some of the important risks that a wheat producer faces? Why? How can those risks be managed?

Fundamentals of Price Risk

The objective of this chapter is to introduce the concept of risk and the process of handling risk.

OVERVIEW

Albert Einstein once said, "I refuse to believe that God plays dice with the universe." He was referring to a field of physics called quantum mechanics where the old certain laws of Newtonian physics were being challenged by new uncertain laws. Dr. Einstein died in 1955 still believing that the new uncertain physics would evolve into certain laws and the ambiguity would be resolved. But, alas, it has not. To date the world of subatomic physics is still ruled by uncertain laws that must be estimated, not calculated with certainty. Yet, modern physics has learned to handle the uncertainty in a way that makes the profession more robust then ever before.

Uncertainty versus Risk

Uncertainty is not a bad thing, as Einstein believed, but simply, as later scientists have proven, something that can be managed when it is identified as a **risk.** The uncertainty that exists in quantum mechanics is not a risk for the average human—it becomes a risk only to the scientists that work in the field—just as the uncertainty of train arrivals and departures in New York is of little concern (a risk) to someone living in California. Thus the difference between uncertainty and risk lies in the impact of the outcomes. *Uncertainty is an unknown outcome. When the unknown outcome has an impact on a person or business it becomes a risk.* Usually the outcome to a person is the result of a conscious decision, but it could also be the result of a passive decision as well.

Hailstorms have uncertainty. The outcome is unknown. A hailstorm may or may not strike a particular farmer's field. But hailstorms are also classified as a risk because they have an impact on people and businesses. A farmer does not actively choose to participate in a hailstorm, but the fact that he lives in a potential hailstorm area makes the uncertainty a risk. It could be argued that the farmer made a choice to live and work in a hailstorm area and it was a conscious decision. But what if the farmer was born and reared in the area? Of course the decision to stay was the farmer's, yet the issue is more complex. The farmer, because of where he farms, is subject to the risk of hailstorms. Whether or not the decision was direct or conscious is really immaterial. All that matters is that the effects of uncertainty bear upon a person or business. It is easy to see that a farmer's direct action to buy a farm in a hailstorm area has made the uncertainty of hailstorms a risk to the farm and the decision to buy a farm in a non-hailstorm area has removed the risk. Direct actions do convert uncertain events into risks for the person or business and will certainly be the dominant form of risk management. Still, it is important to consider indirect or passive actions as well. Again, what matters the most is that the effects of uncertainty have an impact on the person or business.

Now consider the situation of uncertainty where it doesn't directly bear on a person. Producing cotton is an uncertain process. The outcome is unknown. A cotton producer thus has a risk. Nevertheless, the uncertainty of cotton production has no direct impact on a corn producer. Therefore, risks occur when either by direct action or circumstances an uncertain event or process impacts a person or business.

The crux of price risk will usually pertain to a direct action; nevertheless it also can result from the circumstance of the business. A cow-calf producer has the direct risk of price movements for weaned calves. The cow-calf operator is a rancher who does not normally have to feed any supplemental grains to her herd so the uncertainty of grain prices is not a risk to the rancher. A drought occurs and the rancher is forced to buy supplemental corn for her herd; now the uncertainty of corn prices is a risk to the rancher. The rancher has the direct risk of cattle prices and the circumstantial risk of corn prices.

Measuring Price Risk

Risk is a difficult and troubling concept in and of itself and is even more difficult to measure. Over the years, a set of reasonable tools has evolved that are widely accepted as standard procedures to measure risk. Since risk stems from uncertainty, a measure of uncertainty is the first order of business.

Probability and Random Variables

The quantitative measure of uncertainty is *probability*. Mathematically probability values range from zero to one. Zero probability means an uncertain event will never occur and a probability of one means it will happen with certainty. If probabilities are generated from events that have known values, such as a deck of cards, then the probabilities are said to be *objective* because over many trials the probability that a Jack will be dealt is a known number (4/52—4 Jacks in a deck of 52 cards). However, many probabilities are less certain and are said to be *subjective*. What is the probability, for example, that a customer will go to the right when they enter a supermarket? Neither type of probability is necessarily more correct than the other when used to make a decision, but objective probabilities generally give more confidence. Consider for example the objective probability of flipping a coin. Deductively, as well as empirically (by test), we know that the probability that a head will occur on the flip of a fair coin is 1/2 and that a tail will occur is also 1/2. Everyone has confidence in the probability being correct, but that knowledge still doesn't help with what the actual coin toss will be for a single event. By the same token, if the probability that customers will go to the right when they enter a supermarket is subjectively measured to be 5/8 (approximately 63 percent), what an individual will do when he or she enters the store is still unknown. Yet the subjective habit of customers is converted to a probability and used in decisions about how to place certain items within the store. What matters the most is the ability to attach a probability to an outcome, be it objective or subjective.

Table 2–1   Probability Distribution of the Random Variable (Corn Prices)

x = Price of Corn	P(x) = Probability of x
$2.00	0.4
$2.50	0.5
$2.80	0.1
	1.00

The price that a farmer will receive for her corn crop is a *random variable* because the outcome is uncertain. A random variable is simply a numeric value that occurs by chance, i.e., a price. Assigning a probability to each potential outcome would yield a *probability distribution of the random variable* (corn prices), as shown in Table 2–1.

What price would a farmer expect to receive, given the distribution in Table 2–1? The *expected value* would be the sum of the probability of each price occurring times the price, as:

$$E\,(x) = \mu = \Sigma x P\,(x)$$

where

$E\,(x)$ is expected value of x (the *mean*, as denoted by the Greek letter μ)

x = Random variable (price of corn)

$P\,(x)$ = Probability of x

Σ = Sum of all values of x

For the values in Table 2–1, the formula yields:

$$E\,(x) = \$2.00\,(0.4) + \$2.50\,(0.5) + \$2.80\,(0.1) = \$2.33$$

The expected value of corn is $2.33 per bushel given the probability distribution of the corn prices. In other words, over time the expected value that a farmer would receive for his corn is $2.33 per bushel because 40 percent of the time he would receive $2.00 per bushel, 50 percent of the time he would get $2.50 per bushel, and only 10 percent of the time would he receive $2.80 per bushel. The expected value is the average, or mean of the distribution. Figure 2–1 displays the distribution.

Consider the situation where there are 13 corn prices as random variables as shown in Figure 2–2. The expected value (weighted mean) is $2.30. Exactly half of the values lie above $2.30 and half below with matching probabilities of occurring. What happens if an unusually high price occurs, say $4.50, instead of $2.90? The probability is 0.025 that a price of $4.50 will occur, so the expected value (weighted mean) jumps up to $2.34 as illustrated in Figure 2–3. The unusually high price of $4.50 would be termed an *outlier* and may or may not be important. Outliers impact the mean. Another parameter, the *median*, is not impacted by outlier values. The median is simply the value such that half of the observations lie below it and half above it. In the example with the outlier of $4.50, the median would remain at $2.30, while the mean changes as the outlier value changes. Nevertheless, the mean value is the parameter of choice for most risk measurements, unless a good case can be made to use the median.

Once that distribution is completed then the other major parameter of that distribution can be calculated—the **variance** (denoted by the Greek letter σ, squared, as σ^2) and square

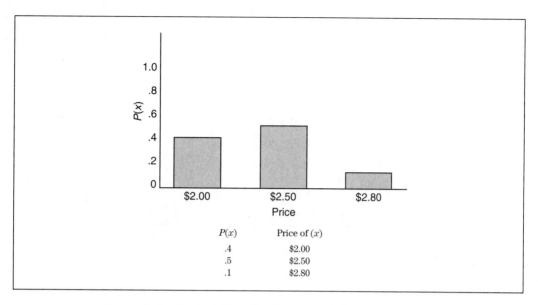

Figure 2–1   Probability distribution of three corn prices

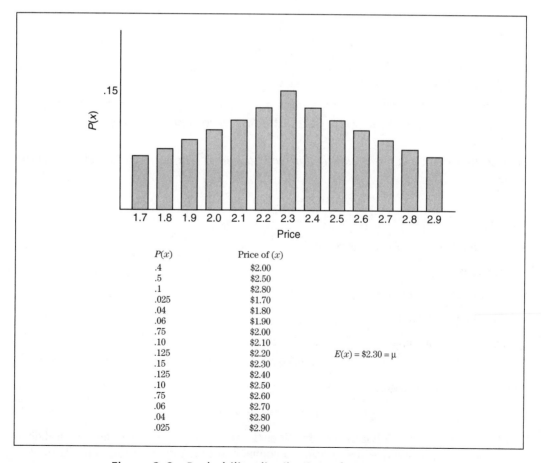

Figure 2–2   Probability distribution of 13 corn prices

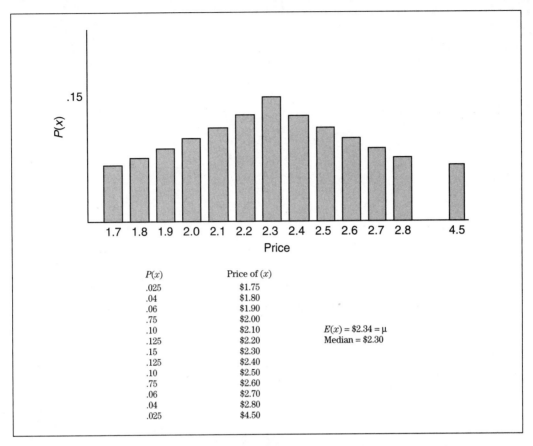

$P(x)$	Price of (x)
.025	$1.75
.04	$1.80
.06	$1.90
.75	$2.00
.10	$2.10
.125	$2.20
.15	$2.30
.125	$2.40
.10	$2.50
.75	$2.60
.06	$2.70
.04	$2.80
.025	$4.50

$E(x) = \$2.34 = \mu$
Median = $2.30

Figure 2–3   Probability distribution of 13 corn prices with outlier

root of the variance–the *standard deviation* ($\sqrt{\sigma^2}$, or σ). The formula for calculating the variance is

$$\text{variance} = \sigma = \Sigma \, (x - \mu^2 P) \, (x)$$

where

     x = Random variable (price or corn)

     μ = mean

     $P(x)$ = Probability of x

     Σ = Sum of all values of x

Using the information found in Figure 2–1, the variance would be

$$(2.00 - 2.43) \times 0.4 + (2.5 - 2.43) \times 0.5 + (2.8 - 2.43) \times 0.1 = 0.0901 = \sigma^2$$

and the standard deviation would be

$$\sigma = \sqrt{0.0901} = 0.30$$

For the data in Figure 2–2 with 13 observations the variance is 0.2087 and the standard deviation is 0.4568. Yet, when a single outlier becomes part of the distribution as in Figure 2–3, the variance jumps to 0.3207 and the standard deviation to 0.57. The standard deviation becomes a powerful risk measurement tool because it is the summary quantity that shows how far from the expected value (mean) the random variable numbers lie (i.e., corn prices in our examples). Thus, the higher the standard deviation, the further from the mean corn price the actual corn prices really are. A low standard deviation implies the actual corn prices are clustered near the average price.

If corn prices have a high variance, they are widely distributed around a central value and thus would be more uncertain and thus more risky to the farmer. In other words, the farmer would have little confidence that a certain price would occur. However, standard deviation is the preferred summary number instead of variance because to calculate variance a squaring process is used to remove negative numbers. By taking the square root of the variance, standard deviation moves the summary number back into the same unit measurement as the actual data. Because variance and standard deviation are so tightly bound, the terms are used interchangeably by risk managers and only need to be differentiated when actual calculations are necessary.

If a couple of added concepts are considered, the major tools of risk measurement are complete. The previous examples contained *discrete random variables*. That is, corn prices were individual whole numbers. For example, in Figure 2–2 corn prices ranged from $1.70 to $2.90 with 10-cent intervals between each price. In reality, corn prices can and do take on any value such as $1.71¾ or $2.31¼, or are more correctly stated as being a *continuous random variable*. A continuous random variable coupled with a large number of observations over a sufficiently long enough time produces a distribution of random variables with unique characteristics. This unique distribution is called a *normal distribution* (also known as a *Gaussian distribution*) and is based on having a continuous random variable and lots of data. The lots of data idea is called the Central Limit Theorem or the Law of Large Numbers, which is really the idea that after a certain amount of information is analyzed, the summary values for the mean and standard deviations begin to behave in a certain, predictable way as expressed in a normal distribution.

A normal distribution is also called the *bell curve* because of the way the distribution looks, as shown in Figure 2–4. The area under the curve represents probability and thus the total area must sum to one. The normal distribution is symmetrical and thus the expected value (mean) is also the median and is defined by the highest point on the distribution.

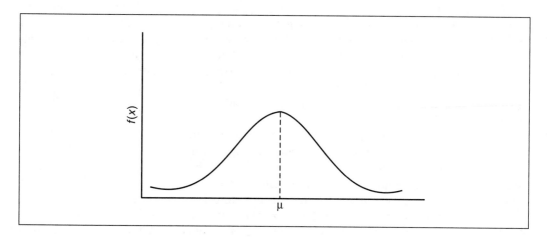

Figure 2–4   A normal distribution or a bell curve

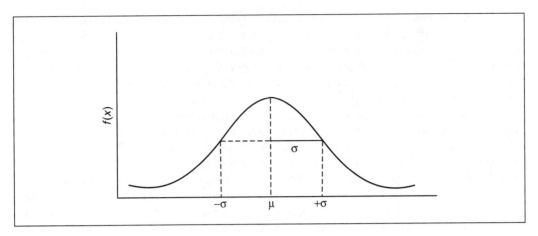

Figure 2–5   Normal distribution defined by μ and σ

As each side of the bell curve begins to quit increasing at an increasing rate and starts to increase at a decreasing rate (the inflection point), that point represents one standard deviation from the mean as illustrated in Figure 2–5. The power of using a normal distribution is that the two summary values—the mean and standard deviations—allow for the specification of each continuous random variable, or more importantly a grouping of random variables. All normal distributions will have these characteristics:

1. 68.3% of all random variables fall within $\mu \pm \sigma$

2. 95.4% of all random variables fall within $\mu \pm 2\sigma$

3. 99.7% of all random variables fall within $\mu \pm 3\sigma$

4. 99.994% of all random variables fall within $\mu \pm 4\sigma$

5. 99.99994% of all random variables fall within $\mu \pm 5\sigma$

6. 99.9999997% of all random variables fall within $\mu \pm 6\sigma$

A continuous distribution will asymptotically approach the axis, but never quite get there. However, after four standard deviations from the mean, almost all values have been accounted. Six standard deviations from the mean are listed to illustrate a modern concept in manufacturing called Six Sigma. The Greek symbol sigma is used to represent standard deviation and its square variance. Pushing out to six standard deviations means very few random variables are not in the distribution. The concept of Six Sigma is to push the quality of manufacturing to the point that only a few defects occur within millions of production units.

Figure 2–6 shows two normal distributions that have the same variance but different means, and Figure 2–7 illustrates the same mean but with two different variances. Figure 2–6 has two different mean values with identical variability. If the mean values of the two distributions represent different corn price possibilities, a rational individual would select the higher average mean possibility since they have identical variances. However, Figure 2–7 presents a different problem. The mean values are identical, but one distribution is less variable than the other. Now the decision must be based upon a person's attitude about accepting higher or lower levels of variability. Those levels of variability really translate to risk.

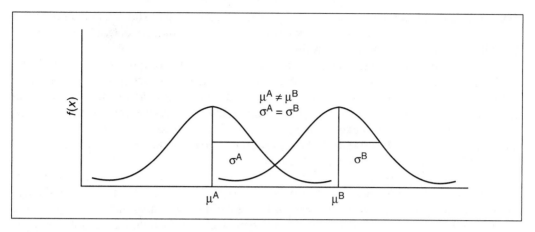

Figure 2–6   Two normal distributions with identical variances but different means

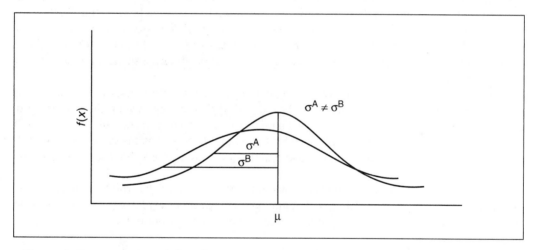

Figure 2–7   Two normal distributions with identical means but different variances

This variability becomes the standard measure of risk–the higher the variance the higher the degree of risk, and vice versa.

Price Risk Management

Management is an active process. It incorporates an action-consequence thought procedure. If this action is taken, what are the likely consequences? Managing price risks includes three major initial actions—**acceptance, neutralizing,** and **transference.**

Acceptance

The acceptance of a price risk is to absorb the full consequence of the uncertainty of the action. It is, surprisingly, the most common form of price risk management—with a twist. Often a price risk is accepted naively. In other words, the manager doesn't know the full consequences of the action (or lack thereof). These are the people who can benefit the most from learning how to manage price risks. However, many managers actively accept price risk with full knowledge of the uncertainty of the action.

Acceptance of a price risk as a price risk management tool consequently has to be divided into two categories—naive and active. If someone naively accepts a price risk, that action is not management. The action was taken without knowledge of the uncertainty of the outcomes. It is, sadly, all too common. If someone actively accepts a price risk, that implies they have full knowledge of the uncertainty of the outcomes. It does not mean they know the outcomes with certainty, but only that they know the uncertain nature of the outcomes and are willing to accept all of them.

Most states have mandatory car insurance laws. You must show proof of insurance or financial means to meet certain levels of monetary obligations. Drivers often find ways around the law and drive without insurance. This is a conscious, albeit naïve, acceptance of the risk of driving. The potential financial consequences of this uncertain action are huge, but may not be known to someone who naively accepts the risk. What about someone that takes the uncertainty of driving and estimates the potential worst case scenario of the pecuniary outcome and decides that they have a large enough economic base to handle the outcomes, should they occur? They actively accept the risk. Most states have provisions for people with sufficient means to self-insure, that is, to actively accept the risk, drive without a formal auto liability insurance policy, and post a bond or other assets as a pledge against potential damages.

Price risk acceptance is common among many businesses. Products are purchased and later sold or sold first and then later purchased with the time lag between the actions fully exposing the business to the risk of price change. The uncertainty of the outcomes are well known and actively accepted. Farmers who have been in business many years know full well the uncertainty of price movements and actively accept the risk that they have a disaster or a boom.

The major issue concerning the active acceptance of a price risk is the knowledge of the uncertainty of the outcomes. All too often managers may think they have a good knowledge of potential outcomes only to be fooled by excessively harsh or boom markets.

Neutralizing

To neutralize a price risk is to remove it completely. Businesses neutralize price risk in three major ways: **forward contracts, passing,** or **tandem actions.** When a corn producer signs a contract to sell corn at harvest for a certain price, he has neutralized the price risk. The price risk has been removed. Passing is the procedure of taking a price risk and passing it on to the next participant. Tandem actions involve two processes that when linked together neutralize a price risk.

Forward contracts neutralize market price risk. The uncertainty of the outcomes of price movements is replaced by a single price. Some agricultural products have strong contracting opportunities such that a producer simply does not begin the production process without a contract that specifies not only the price received, but how much and what quality must be delivered. Unfortunately, in many parts of the United States and for certain products, contracting opportunities are very limited. While forward contracting can neutralize price risk, other risks are added—default risk of the contracting parties and the production risk that the contracting party cannot meet the contract specifications.

A price risk can also be neutralized by passing the risk to another party. Utilities regularly use this process. Fuel costs such as natural gas or coal change and thus impact the cost of producing electricity. A fuel surcharge is included by most electricity generators in the customer's electric bill to cover the risk of fuel price changes. Local distribution companies that provide natural gas to customers typically buy natural gas and then distribute it to customers letting the customer pay whatever the gas cost the company plus a management fee. Similar programs have existed in poultry production operations. The cost of production is passed forward to the next market channel participant and the producer receives a management fee.

Tandem actions are more complicated. To neutralize the risk of a price increase, a tandem action is taken to mitigate the effect. Banks use tandem actions when they price a loan with a variable rate. As the rates increase or decrease, the bank stands to lose or make money. Banks finance the loan with customers' deposits. For interest-bearing deposits, the amount the bank will pay is likewise a variable rate such that the cost of funds to the bank and the revenue received by the bank from loans floats in tandem. If rates go up, the bank pays more for the deposits, but the revenue from the loan likewise goes up, and vice versa. The bank earns the net difference between the costs and revenues which move in tandem with each other, thus neutralizing the risk of price change.

Transference

Price risks are managed by transferring or shifting the risk of price change in one market to another. Transference is more commonly called hedging. The idea is to mitigate the risk of cash price changes by having another financial position that moves oppositely. A gain in one results in a loss in the other market and vice versa. Futures and options contracts are the most common tools used to transfer the risk of cash prices changing. Both of these tools are discussed fully in later chapters. Simply put, the risk that a grain handler such as an elevator has in the cash market is the uncertainty of wheat price movements once the elevator has purchased and stored the grain. The risk can be transferred or shifted to a financial position in the wheat futures market.

Psychology of Risk Management

It is important for price risk managers to know how they feel about price risks because their attitudes about risk influence the way they will handle the risk. During the last 20 years a new direction has emerged in finance called *behavioral finance* or *behavioral economics*. The idea is simply that a dollar is not always a dollar. Behavioral economics puts greater emphasis on the concept of opportunity costs. Consider a gambler who puts $10 at risk at the craps table. The $10 escalates to over $10,000 in a few hours only to fall back to zero before the night is over. When the gambler is asked how he did, his answer is "I only lost $10." Well, no. The gambler lost $10,000. If that same gambler had purchased a share of stock in a friend's company for $10 and watched as the company grew over the next several years and likewise his investment to $10,000 only to see the company file for bankruptcy, he would no doubt say that they lost $10,000, not $10. In both cases, the only difference was time. The gambler lost $10,000 in both cases, but does not believe that except in one case. The human mind does not view all events the same nor does the mind always view them rationally.

Attitudes toward Risk

Attitudes about risk are generally classified as averse (opposed), neutral, or enthusiast (lover). Someone that is price **risk averse** will attempt to manage the risk while a **risk neutral** person will be indifferent. A **risk lover** will embrace the risk. Individuals may of course have all three of these attitudes—but at different times and for different events. A farmer may be very risk averse for her corn crop because it is her livelihood, but may ride motorcycles at 100 miles per hour over rough roads. She is risk averse professionally but personally a risk lover.

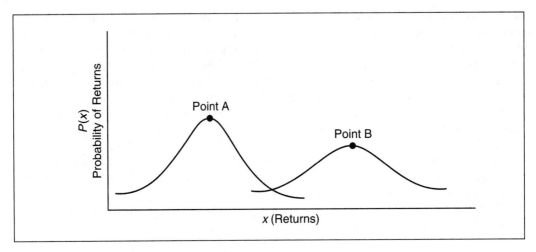

Figure 2–8   A risk-averse decision (Point A) versus a risk-loving decision (Point B)

Precise definitions of each of these three categories has been attempted for years by economists but all fall short because of varying ways of defining risk. Suffice to say, a more general definition is used here as a way to help risk managers attempt to understand their behavior more fully.

A risk averse individual, other things being equal, will opt for a financial gain that has the highest probability of occurring, point A in Figure 2–8. A risk lover would opt for a higher return with a lower probability of occurring, point B in Figure 2–8. A risk neutral individual would be indifferent between any combination of probabilities and returns between points A and B.

If, for example, a wheat producer is risk averse, he would accept a marketing strategy that would yield less but have a higher probability of occurring than a risk lover. A wheat producer who is a risk enthusiast would accept a strategy with a lower probability of occurring so that he might enjoy a higher return. On the other hand, given a certain level of probability of occurring (point C and point D in Figure 2–9), a rational decision maker, regardless of whether or not he is risk averse or enthusiast, will select the highest return level (point D versus point C in Figure 2–9).

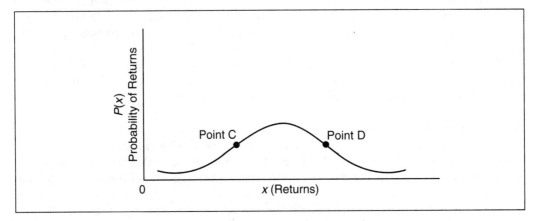

Figure 2–9   Rational risk/return tradeoff

Notwithstanding the previous discussion, the tenets of behavioral economics say that the distinction between risk averse and enthusiast may be blurred. Decisions are not always rational, and if they are, sometimes all of the relevant information is not available.

It is extremely difficult to quantify price risk and even more demanding to quantify an individual's attitude toward risk, but it is important to risk managers to try and understand why they prefer one course of action to another and what the likely consequences will be. As the old saying goes, "It may not be a game worth playing, but it is a game worth winning."

Developing a Risk and Mitigation Profile

To start the process of determining what risks exist in a business situation requires a systematic look at the risks and whether or not those risks can be mitigated and how. This process is called a **Risk and Mitigation Profile** and contains the following elements: (1) What are the risks? (2) Can the risks be mitigated? (3) How can the risks be mitigated? (4) What are the costs and benefits? (5) Do I want to mitigate the risks?

What are the risks? This first step is defining the risks the business is facing. The risks should be classified according to type—policy, production, default, or price, for example. This sounds like a daunting task, but the efforts generally pay off by providing a better understanding of the overall business. However, because many of the risks are risks the business will have to absorb, oftentimes this step is shortened to only price or financial risks.

For example, a corn and soybean farmer certainly has policy and production risks, but the two major price risks are corn and soybean price changes. As a first step, the farmer may want to think about and list as many risks as possible, but later look more closely at just price or financial risks.

Can the risks be mitigated? The second step is the reason that most risks of a business are listed only once in the initial profile because many risks a business faces simply do not have mitigation possibilities. For the corn and soybean producer, the two price risks can be mitigated. Most price and financial risks can be alleviated by some tool.

How can the risks be mitigated? For corn and soybean price risks three major possibilities exist—futures contracts, option contracts, and forward contracts. Additionally, there may be swap opportunities or various forms of different types of forward contracts offered, depending on where the farm is located.

What are the costs and benefits? What does the producer gain if futures contracts are used? What are the brokerage fees and time value of money costs associated with initial margins and potential margin calls? Similarly, what are the brokerage fees and time value of money costs associated with options and forward contracts?

Do I want to mitigate the risks? Deciding whether to mitigate risks is a painful but important step in deciding the farmer's attitude toward risks. If the Risk and Mitigation Profile has been done properly, the farmer now has an excellent set of information to make a more informed decision.

A Risk and Mitigation Profile appears at the onset to be a disheartening task. Nonetheless, once the process has been completed, subsequent profiles become easier and become a failsafe document used with brokers, boards of directors, managers, and others with whom the risk manager must deal concerning the process of managing risks. In later chapters the profile will be merged into formats with examples.

A Final Word about Risk

A wag once said, "I don't know art, but I know what I like." To a large extent risk is similar. It is difficult to define what risk is in general and almost impossible for each individual situation, but people can recognize it and define it for their own individual or business situation. Academics attempt to define risk with mathematical and statistical tools so it can be more readily quantified, but statistical measures such as variance and covariance may simplify the risk an individual faces too much and may miss entirely because past data and assumptions don't work anymore.

Yet the process of trying to understand risk more fully is worth the effort, just as it is worth the effort as an individual to try to determine psychologically why you have certain attitudes toward risk and how to handle those risks and attitudes. It is by no means a perfect process, but it can yield powerful results for businesses. Modern physics has moved forward despite the uncertainty found at the subatomic level. In fact, the Nobel Prize was awarded to Werner Heisenberg for his Principle of Uncertainty, which defined the limits for the new physics and prompted Albert Einstein's comment about God playing dice with the universe.

It appears that God does in fact play dice with just about everything in the universe forcing uncertainty on all situations, especially financial ones. The more we attempt to understand the uncertainty, the better we can more fully manage risk—certainly a game worth winning.

CHAPTER 2—QUESTIONS

1. Tom owns a small ranch that has 200 cows. He sells the calf crop each year after weaning. What kind of Risk Profile would Tom have?

2. Is risk the same as uncertainty?

3. What is the major difference between neutralizing a risk and transference of a risk?

4. How could a corn producer use tandem actions to manage the risk of corn prices going down?

5. Should a risk lover ever consider using risk management?

Price Forecasting

KEY TERMS		
blowing up	fundamental analysis	candlestick
Efficient Market Hypothesis (EMH)	price elasticity of supply	curve fitting
Inefficient Market Hypothesis (IMH)	price elasticity of demand	moving average
Random Walk Hypothesis (RWH)	bar chart	oscillator
technical analysis	point and figure chart	

This chapter will help the reader understand the major elements of price forecasting.

OVERVIEW

The truth is simply this: no one can accurately forecast any market. An old bit of wisdom says, "He who forecasts the future, lies. Even when he tells the truth." No one knows with certainty what the future holds, so any forecast is a lie; and even if a future event were correctly forecast, the forecaster didn't know positively that the event would occur, thus still a lie. Lie might seem a harsh word—no harm or any malice is intended. And yet, all forecasters know conditions change as soon as they make their forecasts. Thus, forecasters always tell you something they know will not happen exactly as they say it. A wise trader understands this and proceeds into the world of price forecasting as if it were a mine field. The object of getting across a minefield is to do so with out getting blown up. Not so coincidently, the expression used by traders to signify that they have lost more than they can stand is to say that they "had a **blow up**" or merely "**blew up.**" In reality, traders can blow up for many reasons, but betting on a price direction that was wrong is a common mistake. What to do then? If a market price cannot be correctly forecasted, why do people constantly try? The short answer is that being on the right side of a price move has value. The quest for money causes people to forecast prices regardless of whether or not they are accurate.

This chapter is not designed to be a primer on how to forecast prices, but rather a discussion on the various ways that prices are forecasted so that risk managers can make more informed decisions on which risk management tools are more appropriate for their individual needs. Additionally, it is important for all risk managers to develop their own philosophy concerning how markets behave.

The Two Big Mine Fields

Traders believe that prices can be forecast or they cannot. Those who maintain that prices cannot be forecast believe in the **Efficient Market Hypothesis (EMH)** and those who believe otherwise are proponents of the **Inefficient Market Hypothesis (IMH)** (also called the Deterministic Hypothesis). The EMH says that the mines in the field have been randomly distributed and that out of all the traders who enter the field, a certain number on average will emerge and all of the others will blow up. No individual trader can possibly figure a way through. IMH declares that while an individual mine's location may be unknown, patterns and certain causes and effects can be known and a route through the field can be drawn with fewer blowups. Mines, for example, may have been placed in greater concentrations in the middle and thus the edges of the field are safer. Or left-handed people set mines and therefore the pattern can be skewed to the left. IMH maintains that with enough study and information, a better way can be found through the minefield.

Efficient Market Hypothesis

Louis Bachelier stated in 1900, "The mathematical expectation of the speculator is zero."[1] Bachelier's idea was termed the **Random Walk Hypothesis (RWH)** from an earlier discussion among scholars. The question was posed, "If you leave a drunk friend in a garden that is enclosed with a locked gate, when you arrive the next day to pick him up, where will the most likely spot be?" The answer early scholars arrived at was in the spot where you left him. If he was drunk then each step he took would be random and thus the expected result of his steps over some time would sum to zero and thus the most likely mathematical spot to find him is where he was left, ergo, the name Random Walk Hypothesis. RWH evolved in the 1960s to the EMH when Bachelier's earlier work was rediscovered and new research added. An efficient market has large numbers of traders who use all available information and all future expected information to formulate price. Since price has all known and knowable information embedded in it, including all random news as it occurs, it will be unrelated to any other price. Over the years, the EMH has been codified into three major forms:

- **Weak Form**—All past information is reflected in price discovery.

- **Semi-Strong Form**—All past information as well as all current known information is used to formulate prices.

- **Strong Form**—All past and current information plus all knowable information is considered in the pricing process.

The price forecasting literature has been filled with countless articles and research endeavors that support the EMH since the 1970s. The three forms have evolved to reflect the various beliefs among traders on the validity of the EMH. Weak Form believers think the markets are generally efficient, but not all the available information is fully incorporated into the pricing process. The Semi-Strong Form states that all useful information, past and present, has been used and only insider information could change the market price. The Strong Form advocates state that even insider information has somehow been embedded in the market price.

Inefficient Market Hypothesis

IMH is the theory that market prices are not determined with perfect information and instead are constantly evolving as more information becomes available and is used by traders. IMH champions believe that with knowledge and skill some measure of price forecasting can be valuable as market inefficiencies are discovered and acted upon. How inefficiencies are

---

[1]Bachelier, L. (1900). "Thèorie de la Spìculabon" (The Theory of Speculation), *Annales de l'Ecole Normale supériure.*

discovered fosters almost as many fierce arguments among believers as between the IMH devotees versus the EMH supporters. IMH adherents are broadly classified as either technical or fundamental analysts.

Technical analysis is a method of analysis based on the belief that where the market has been in the past is, in some way, the best indicator of where it will be going in the future. Technical analysts dismiss the Weak Form of the EMH. Technical analysis is divided into two categories: charting and mathematical modeling.

Fundamental analysis holds that price determination has a cause-and-effect relationship and once the cause is properly identified, the effects can be forecast with some degree of accuracy. Fundamental analysis uses economic data and relationships, knowledge about events and circumstances, and any other data or causal connections to ascertain price. Fundamental analysis rebuffs the Semi-Strong Form of the EMH.

Perhaps another form of IMH followers should be identified—insider trades. The Strong Form of the EMH states that not even insider information alters the random nature of price discovery because it has already been anticipated. IMH disciples dismiss the Strong Form of the EMH as rubbish. Insider information does move markets, the IMH folks argue, and is unknowable in advance so cannot possibly be incorporated in the price. EMH proponents counter that even though insider information actions may be unknowable, the effect on the market price over some time period is insignificant. Since insider information cannot be used legally to trade, this aspect of price forecasting will be left untouched.

Using the Efficient Market Hypothesis

If markets are efficient and therefore cannot be forecast, of what use is the theory? Certainly the EMH is not used to forecast prices, but if markets move efficiently then certain other aspects of the market price movement might have value to risk managers.

The broadest use of the EMH by traders occurs in the equity markets. EMH believers do not believe that active mutual fund managers are any better than a simple index of various stocks such as the Standard and Poor's Index of 500 stocks. An equivalent idea in agricultural markets is corn price in Iowa. The EMH model asserts that the price of corn in Des Moines is efficient and cannot be forecast with any accuracy. However, that does not mean that certain characteristics of the price cannot be helpful to a trader in the short run.

Next Day Pricing—What is the most likely price of corn tomorrow? The best guess is not a random number pulled from all possible numbers, but rather what the price is at the end of the previous day. Where the price was yesterday is a better guess for where it will be today than a simple random guess. We suspect today's prices will not be the same as yesterday's and yet we do not think they will be vastly different. If Des Moines corn price yesterday was $3 per bushel, the best guess for today is $3 per bushel rather than a guess that doesn't acknowledge or relate to the general level of prices in the recent past. On the other hand, a guess of today's price based on yesterday's price will be no more accurate according to the EMH than a guess of say $3.20 per bushel or $2.80 per bushel if the forecasts are based on the relative level of yesterday's price.

Short-Run Minimum/Maximum Prices—Where the market has made a new high or low in the short run is a better guide for the short-run maximum and minimum price forecasts than a simple random guess that ignores general price levels. For example, if, during the last three months, Des Moines corn price reached a high of $3.30 per bushel and a low of $2.75 per bushel, those two prices would be a better guide for potential highs and lows for the immediate futures than a simply random guess. The EMH says that this would, however, not be true for a longer period of time.

EMH believers will not try to second-guess where the market price is headed, but rather will use past short-run price movements only as a guide for general price level expectations. If, for example, a manager needed to value a bin of corn for inventory purposes that will be sold in two weeks, what today's market price is or last week's average is a better guide for valuing the corn now than trying to forecast what corn prices will be in two weeks, because accurate forecasting is impossible and thus the general level of prices now is the best price to use.

Using the Inefficient Market Hypothesis

Obviously the idea that prices cannot be forecast with any accuracy doesn't appeal to a lot of traders. The EMH appeals to a lot of academic researchers but has little appeal to the general trading population. Why else would they be trading? Traders are always looking for an edge and they use many techniques to forecast price direction.

Fundamental Price Forecasting

One of the major bodies of economic theory is concerned with supply and demand and the interaction between the two to determine price. Alfred Marshall first proposed the idea in the early 1900s that equilibrium price was the result of the scissor-like connection between a supply curve and a demand curve as shown in Figure 3–1.

Supply
Modern microeconomic theory says that an individual producer's supply response curve will be the upward sloping portion of his marginal cost curve and the market supply curve will be the horizontal formation of all individual marginal cost curves. Marginal cost is the cost

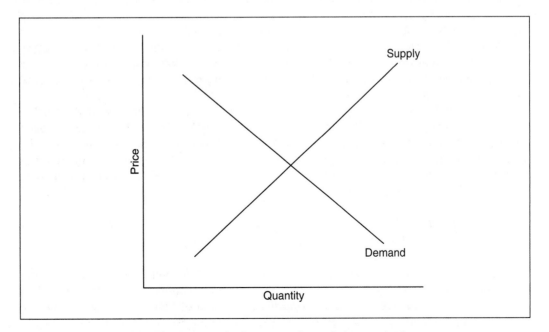

Figure 3–1 Marshallian supply and demand scissors

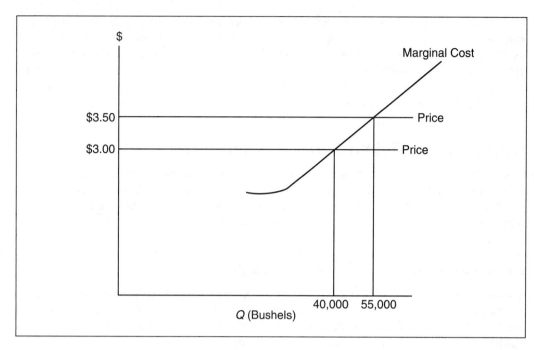

Figure 3–2   Marginal cost curve for a producer

of the next unit of production as shown in Figure 3–2. An individual producer will produce at a point where the price received is equal to the last unit produced, implying that all previous units had a cost that was lower than the price. To induce a producer to produce more, the price must increase, thus the marginal cost curve is also the individual's supply response curve. This wheat producer would produce 40,000 bushels at a price of $3 per bushel, but if the price were $3.50 per bushel he would produce 55,000 bushels.

Figure 3–3 shows the hypothetical supply curve for wheat, which would be the **aggregation** of the wheat producers in the United States. The market supply curve traces the quantity that will be supplied as the price of wheat changes. Thus price changes will cause a change in the quantity supplied.

How much the quantity supplied will change due to a change in price is called the **price elasticity of supply.** If the price changes a great deal and the quantity supplied doesn't change much, the supply curve is said to be inelastic. If the price changes a little and the quantity supplied changes a lot, the curve is said to be elastic. Price elasticity of supply is equal to the percent change in the quantity due to a percent change in the price or over some range from Q_1 to Q_2. The formula is

$$E = \frac{\text{Percent change in quantity supplied}}{\text{Percent change in price}} = \frac{Q_2 - Q_1}{(Q_1 + Q_2)/2} \Bigg/ \frac{P_2 - P_1}{(P_1 + P_2)/2}$$

If a price change of 1 percent causes more (less) than a 1 percent change in quantity supplied, then the curve is said to be elastic (inelastic). How responsive a quantity supplied change is due to a change in price is primarily a function of how fixed or inflexible the resources or production processes are relative to a time period. The biological process to produce most grain crops is fixed. If the price of wheat goes up a great deal in December, wheat

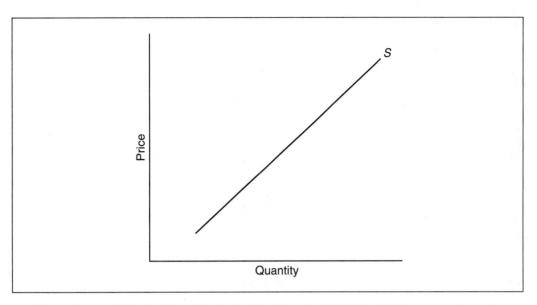

Figure 3–3   Hypothetical supply curve for wheat

producers cannot produce more wheat and get it to the market simply because the biological process of planting and growing wheat is determined by the season of growth. However, if the price of cheese goes up, cheese plants could respond more quickly by putting on extra shifts or adding more equipment. The wheat supply curve would be relatively inelastic (unresponsive to price change) relative to the cheese supply curve, as illustrated in Figure 3–4.

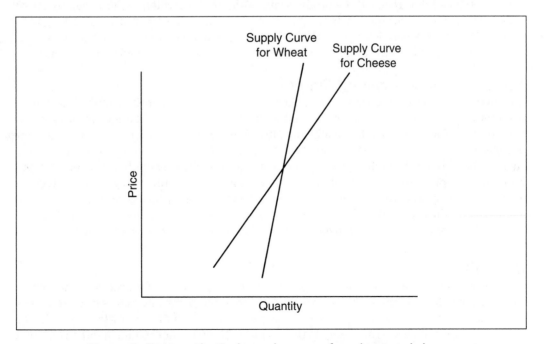

Figure 3–4   Hypothetical supply curves for wheat and cheese

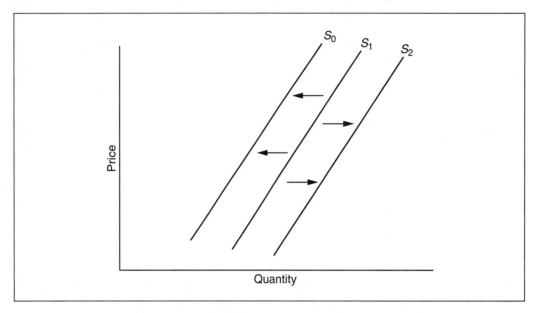

Figure 3–5   Hypothetical supply curve changes

A market supply curve exists at a given point in time at which production technology is relatively fixed and the price of inputs used to produce the product is fixed. If either of these items changes, then a whole new supply curve will exist. It is said that supply will change when production technology or input prices change. If a new production technology becomes available that will be more productive then the supply curve will shift to the right. If a pesticide was banned that reduced the productivity of the wheat farmer, then the supply curve would shift left. Likewise for a change in the price of inputs. If fertilizer prices increase, then the supply curve will shift left. Figure 3–5 shows some hypothetical supply curve changes. In Figure 3–5, S_1 represents the original curve while S_0 shows a decrease and S_2 an increase.

Major Points Concerning Supply

A change in the price of a product will cause a change in the quantity supplied, and how responsive that change is will determine the elasticity of supply. Changes in production technology or input prices will result in an entirely new supply curve that is either to the left or right of the old one. Also, since the market supply curve is a sum of all individual supply curves, changes in the number of producers will shift the market supply curve left or right. Fundamental price forecasters will concentrate on watching for changes in production technology, changes in prices of major inputs in the production process, and changes in the number of producers. They will also attempt to know the price elasticity of the supply curve so they can estimate the change in quantity supplied when a price change occurs.

Demand

Demand curves are derived from consumers' utility (value/use) of a product. The idea is called Diminishing Marginal Utility. One unit of bread has a certain value or use to a consumer. However, a second unit of bread will in most cases result in somewhat less value or use and a third even less. The theory is that to induce a consumer to take another unit of a product that clearly has less value than the previous one, the price of the product must be

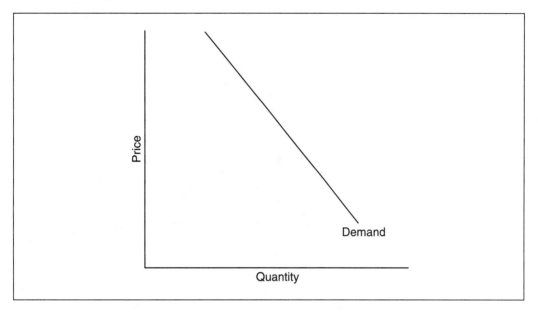

Figure 3–6   Hypothetical market demand curve

lowered. Consequently, an individual's demand curve slopes downward to the right. The market demand curve is the sum of all individual demand curves and slopes downward to the right just like individual demand curves as shown in Figure 3–6.

Price elasticity of demand describes how responsive quantity changes are to changes in price. Price elasticity of demand is equal to the percent change in quantity due to a present change in price, or over some range of Q_1 to Q_2. The formula is

$$E_D = \frac{\text{Percent change in quantity}}{\text{Percent change in price}} = \frac{Q_2 - Q_1}{(Q_1 + Q_2)/2} \bigg/ \frac{P_2 - P_1}{(P_1 + P_2)/2}$$

If price changes 1 percent and quantity demanded changes by more than 1 percent, the product is said to be elastic and if the quantity demanded changes by less than 1 percent it is inelastic. The degree of elasticity of a product is roughly determined by the amount and availability of substitutes for the product. Items that have many substitutes tend to have an elastic demand because if the price changes, consumers can readily substitute other items. Products that have few, if any, substitutes have inelastic demand curves, whereas if a major price change occurs, consumers cannot adjust their consumption very much because of the lack of choices. Price elasticity of demand is fickle. What is a substitute to one consumer is not to another. To a consumer who is rich or simply brand loyal, a certain type of luxury car may have few if any substitutes and therefore have a fairly inelastic demand curve that otherwise would be elastic for another group of consumers.

The individual's demand curve is derived by holding income, tastes and preferences, and the price of other goods constant. When the sum of all individual demands creates the market demand, then the population of consumers is likewise fixed. If income, population, tastes and preferences, or the price of other goods change, then the market demands will shift—left for a decrease and right for an increase. If new health information reveals that red wine improves health, then peoples' tastes and preferences concerning red wine might

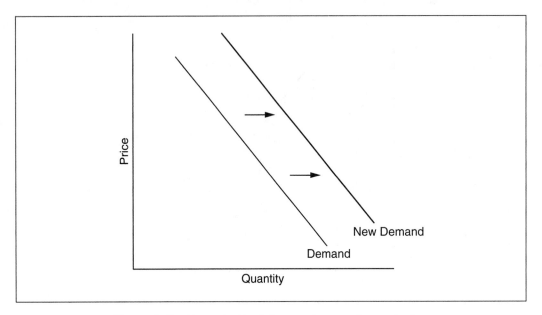

Figure 3–7 Hypothetical demand curve for red wine

change and shift the demand right, ergo an increase in the demand for red wine as illustrated in Figure 3–7.

Major Points Concerning Demand

Price changes cause a change in quantity demanded and the amount and availability of substitutes will determine how responsive the change in quantity demanded will be. A whole new market demand curve will exist if population, income, tastes and preferences, or the price of other goods change. Fundamental price forecasters will endeavor to know how price elastic the demand curve is for a product so they can more correctly estimate changes in quantity demanded when price changes. They will also estimate changes in income, population growth, tastes and preferences, and the price of other goods. For example, if the price of pork changes, then the demand for beef will be affected since beef and pork are substitutes for many (but not all) consumers.

The relationship between the price change of one commodity and what it does to the quantity demanded of another product is measured by *cross price elasticity of demand*. The formula is

$$E_X = \frac{\text{Percent change in quantity of A}}{\text{Percent change in price of B}} = \frac{Q_{2A} - Q_{1A}}{(Q_{1A} + Q_{2A})/2} \bigg/ \frac{P_{2B} - P_{1B}}{(P_{1B} + P_{2B})/2}$$

If a 1 percent increase in the price of one product induces a positive change in the quantity demanded of another product, the two products are said to be *substitutes* (such as beef and pork). If, on the other hand, a 1 percent increase in the price of one product causes a negative change in the quantity demanded of another product, the two products are said to be *complements* (such as beer and pretzels).

Additionally, as consumers get older, their life-long consumption patterns don't change very much. That is, if someone grew up consuming bacon and eggs for breakfast, they will likely continue that trend over time. However, the consumption of bacon and eggs might slowly change due to health or weight concerns as the consumer ages. Demographic and cohort analysis (a similar group of people) are important fields of study for demand analysis, especially over time.

Putting Supply and Demand Together
Perfect Market Model

A perfectly competitive market is defined as a marketplace with many buyers and sellers who are not large enough to have any undue influence, vying for a homogenous product. Not too many markets meet the criteria of a perfect market. However, many markets are said to be *workably competitive.* A market may have many buyers, for example, but only a handful of sellers, or vice versa. But if the small number of either buyers or sellers can't exert any type of monopoly power, then the results of the not-so-perfect market are similar to a perfect market. A perfect market is one where there is no excess economic profit. Prices reflect costs including a return that keeps the resources employed in that use (i.e., opportunity costs), but no more.

Consider, for example, that the price of wheat in Chicago is $4 per bushel but the price for the same kind of wheat in Dallas is $5 per bushel. The cost of transportation between Chicago and Dallas, or vice versa, for a bushel of wheat is 75 cents per bushel. The perfect market model says that the price difference between the two markets should reflect the cost of transportation, but no more. The current difference between the two markets is $1 per bushel, yet the cost of transportation is only $0.75 per bushel. A rational trader would buy the wheat in Chicago for $4 per bushel, ship it to Dallas and sell it for $5 per bushel, and earn an excess economic profit of 25 cents per bushel. Traders who do this are called *arbitragers.* Arbitrage is the process of capturing excess economic profits between two or more markets. The actions of arbitragers will cause the excess economic profits to vanish. In the example between the price of wheat in Chicago and Dallas, there will be several traders who will try to capitalize on the excess profits. Arbitragers will enter the market in Chicago (an increase in the number of market participants, thus a shift right of the demand curve for wheat) thus bidding up the price of wheat in Chicago. Likewise, arbitragers will be new market suppliers in Dallas at $4.75 per bushel or above (an increase in the population and thus a shift right in the supply curve) in Dallas, thus bidding the price down. Figure 3–8

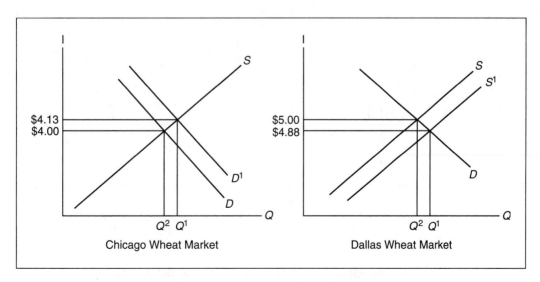

Figure 3–8   Actions of arbitragers that change prices between
two spatial markets that are imperfect

shows these effects with supply and demand diagrams. Markets that are separated by space have *spatial* price differences. Perfect spatial markets differ by the cost of transportation, no more or less, as spatial arbitragers will bid away any excess economic profits.

Markets differ also by time and product form. Time different markets are *temporal* markets and should differ in price by the cost of storage (carry). Products that undergo processing from the raw form to a finished good should differ in price by the cost of processing and are said to be *form* markets. Thus the three types of market models are spatial, temporal, and form. Using the perfect (or workable) competition model, arbitrage traders will access what should be the price difference between each of these markets (transportation costs, storage costs, and processing costs) and determine if excess profits can be made. When they can exploit the differences, they will do so and drive the markets back to normal. Profit taking by arbitragers helps maintain competitive markets.

A futures contract that has a delivery date of December has a price relationship with the current cash price, say in October. If we take away the spatial and form differences and have the product only differ by time, what should be the relationship between the cash price of corn in Chicago in October and the futures price for delivery in December? The perfect market model says that the temporal price differences should be only the cost of storage from October to December (two months). If the price of cash corn in October is $3 per bushel and it costs $0.10 to store it until December, the December futures price should be $3.10 per bushel. If the futures price was $3.15 per bushel and the cash price was $3 per bushel, arbitragers would enter the market. They would sell the December futures contracts (promising to deliver in December) for $3.15 per bushel and simultaneously buy local cash corn for $3 per bushel. They would store the cash corn for two months incurring a cost of carry of 10 cents per bushel and earn 5 cents per bushel excess economic profit. Figure 3–9 illustrates how the action increases the demand for cash corn in October, thus increasing the price while the supply of December delivery corn increases, thus lowering the price in December. Actions of temporal arbitragers keep the futures and cash markets tightly linked via the cost of carry.

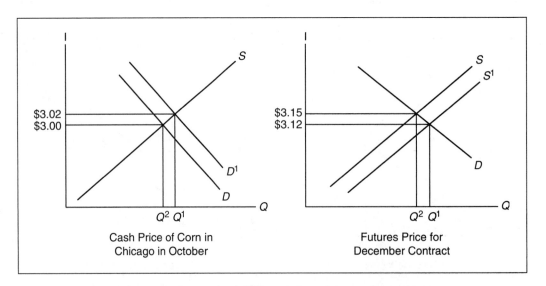

Figure 3–9 Actions of arbitragers that change prices between
the temporal markets that are imperfect

Soybean markets are linked to the finished markets for soybean oil and meal via the cost of processing the raw soybeans into the finished forms. The price difference between the raw product market and the finished product market should be the cost of processing. Form arbitragers will enter the market when there are profit opportunities.

The perfect market model provides a theoretical framework to analyze if markets are inefficient. If spatial, temporal, and form markets differ by the cost of transportation, storage or processing, no more, no less they are said to be *perfect markets* and no excess economic profit exists. If markets have potential profits in them because the price differences between the markets are larger than the costs of transportation, storage, or processing, then they are said to be *imperfect markets* and arbitragers will exploit the excess profits away. If market price differences are less than the cost of transportation, storage or processing whereby a potential loss exists, they are said to be *not perfect markets*. Not perfect markets function slightly differently than imperfect markets. In the previous temporal example in Figure 3–9, arbitragers simultaneously had positions in both the futures and cash markets to capture the excess profit. If, however, the December futures price was $3.07 and the cash price was $3 with cost of storage for two months at $0.10, an action of buying the cash and selling the futures would yield a loss of three cents per bushel. What might happen is traders who needed corn would buy the December futures delivery corn for $3.07 today. If they bought the cash corn today and stored it until December, the cost would be $3.10, thus it would be cheaper to buy the futures contract and wait for delivery in December. Figure 3–10 shows this action by increasing the demand for December corn futures, thus increasing the price and driving the price difference back to the cost of storage of 10 cents per bushel.

Arbitragers will keep spatial, temporal, and form markets closely tied together such that the movement in one market will also impact a related market. Their actions cause derivative market prices and cash market prices to "tend to trend together." Consider, for example, October cash corn prices at $3 per bushel and December corn futures prices at $3.10 per bushel and the cost of storage at 10 cents per bushel from October to December. The two markets are said to be perfect. A major winter storm hits all of the Midwest so

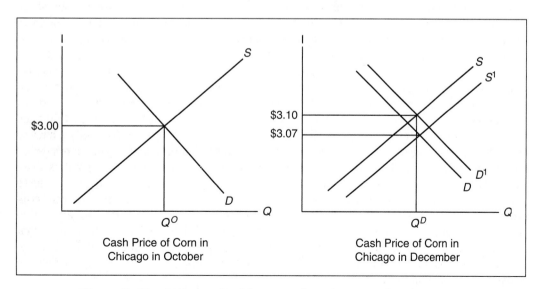

Figure 3–10   Actions of arbitragers that change prices between the temporal markets that are not perfect

severely that cattle have no grass to eat and need supplemental grain to survive. The demand for cash corn will expand, pushing local cash prices up. What would the December corn futures price do? It should move in tandem with the cash price and differ by the ten cents storage cost, that is, trend together. The December corn futures contract derives its value from the cash market and the difference should be the cost of storage.

The Law of One Price

The perfect market model provides the framework for the theory that price differences should be able to be explained by the cost of transportation, storage, or processing—thus only one price plus or minus expected costs. Some refer to the perfect market model as the law of one price. If the difference between two or more prices can be justified by cost, then the two prices are identical except for defensible costs and are said to be one price. The law of one price is also used to explain the differences between identical products that are separated by different currencies. The price of copper in the United States should be the same price as the same grade of copper in another country adjusted for currency differences. If not, currency arbitragers will enter the market. Acting separately in the same market, currency arbitragers and copper arbitragers will force price equilibrium between grades and between currencies. *The Economist* regularly publishes its Big Mac Index that is the cost of a Big Mac hamburger at McDonald's in several countries adjusted for currency differences. To be sure, the price differences are not always reflective of currency differences. But it does provide a reference point to look at other supply and demand factors that might cause a price difference other than currency.

The law of one price is simply another way of looking at the perfect market model. It provides a starting point for arbitragers as they attempt to extract any excess economic profits that may exist in imperfect markets. Arbitragers are the ones that keep markets separated by justified cost differences—transportation, storage, processing, or currency.

Artificial Price Floors

When market supply and demand curves are put together, the result is a market-determined price, or the classic scissor diagram. This model is useful for forecasters to use to look at the effects of circumstances or governmental action that would impose a floor or ceiling on price. In Figure 3–11, a governmental program decrees an artificial price floor. At the nonmarket-determined price, more producers are willing to produce the product than consumers are willing to buy at the artificial price and thus a surplus results. A surplus will cause stockpiles to increase and pressure for "black" markets to emerge where the product trades unofficially at a lower-than-decreed ceiling price. This regularly happens in many U.S. price-supported commodities such as sugar. The U.S. government supports a domestic price that is higher than the world price of sugar, thus there is a constant surplus of sugar on the U.S. market. Price forecasters have to be aware of various governmental price support programs and their likely impact on prices.

If an artificial price is set below market equilibrium, then a shortage will result as more consumers will want the product at the nonmarket lower price than suppliers are willing to provide to the market as shown in Figure 3–12. This occurs in cities that impose rent controls on apartments and in the last few years in the medical field when either governmental actions or actions of Health Management Organizations (HMOs) artificially fix a price that is below the market equilibrium price. It rarely occurs in agricultural commodities in the U.S. but does occur abroad such as in Mexico where corn prices were set very low due to pressure from consumers.

Price forecasters have to be vigilant in addressing the various nonmarket pricing schemes that exist in order to properly determine the likely impact on prices, not only for the domestic U.S. market but also for the international market as well if the commodity is traded globally.

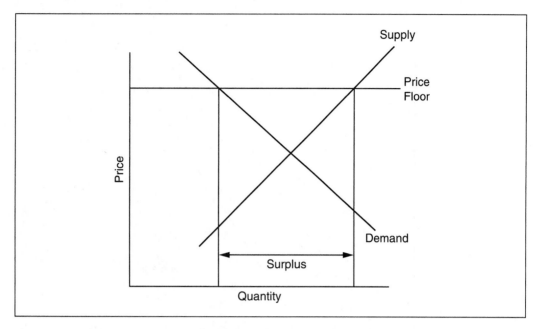

Figure 3–11   Artificial price floor creating a market surplus

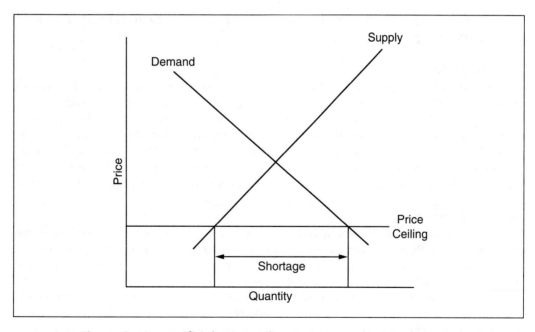

Figure 3–12   Artificial price ceiling creating a market shortage

Supply and Demand Driven Prices

Certain situations exist in a particular market that will result in most of the price movement being caused primarily by either changes in supply or demand, but rarely both. If most of a price movement is caused by changes in supply, the price movement is said to be

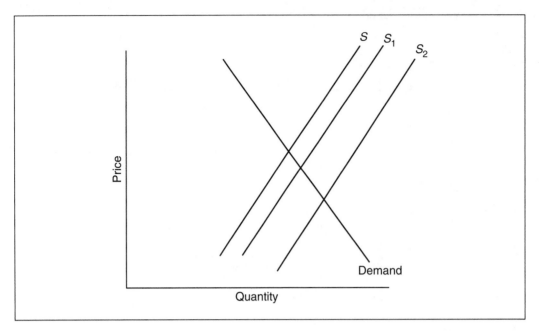

Figure 3–13 Effects of supply changes on the price of corn

supply driven. If the majority of the movement is from changes in demand, it is called demand driven price changes.

Supply Driven

Figure 3–13 shows a hypothetical situation for corn. Demand for corn is fairly stable as uses are well established and will change only by some small percentage as population changes or as new uses are developed. However, the supply curve will shift as growing conditions change causing the price of corn to be influenced more by the supply curve shifting than changes in demand. As the supply curve shifts from S to S_1 and S_2, the price of corn changes. Price forecasters would therefore concentrate on studying the factors that influence the supply of corn such as constant monitoring of weather and growing conditions as reasons for price changes.

Demand Driven

Fine wines are primarily demand driven in terms of price movements. People's tastes and preferences are influenced when they know that the vintage will be excellent. The supply of the wine is fixed at a certain volume and the entire price changes will occur on the demand side as the vintage's quality is assessed by wine consumers, as shown in Figure 3–14. The price of wine will change as the demand curve shifts from D to D_1. Price forecasters will watch for reports from critics, restaurants, and other quality assessors to help determine how demand will change, as supply is fixed for a single vintage wine, and will gradually shift to the left as the vintage is consumed.

Seasonal and Cyclic Movements

Agricultural commodities have biological characteristics that impose a seasonal factor on the production process. Corn is planted in the spring and harvested in the fall in the Northern hemisphere and the opposite in the Southern hemisphere. Hard red winter wheat

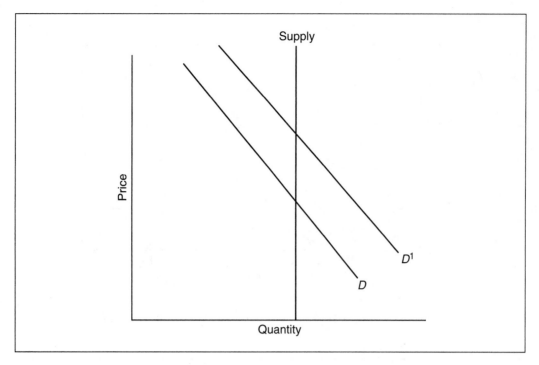

Figure 3–14   Effects of demand changes on the price of wine

is planted in late summer or early fall and harvested in the late spring or early summer of the next year. The nature of seasons and biology also impact animal agriculture as well. Most ranchers calve in the late winter and early spring and then wean the calves in the fall. Each species of animals has different gestation periods and production characteristics and requirements. Therefore, each agricultural commodity will have a different seasonality associated with the production of the product. Seasonal price movements are price activities that occur within a calendar year or production period. Cyclic price movements are price tendencies that occur over several production periods or years.

It is important in price forecasting to understand whether or not a product has a seasonal price tendency or a price cycle. Almost all crops have a seasonal price pattern as shown in Figure 3–15. Some animals show cyclic price behavior as illustrated in Figure 3–16 for cattle.

Econometrics

Two major types of price forecasters use supply and demand analysis: "gut" analysts and econometric analysts. Gut analysts filter all of the various shifters of supply and demand through their brain and come up with a price estimate. It is impossible to teach how to be a gut forecaster. Each individual will learn a particular way and will process information differently. Gut forecasters use the basics of supply and demand and other standard economic models and then come up with a forecast based on experience, judgment, and intuition. Many gut forecasters are very good at what they do; no doubt many are very bad, but they don't last long. This is plainly one area of price forecasting that is difficult to measure and therefore is left to the reader. The other area, econometrics, while complicated and mathematically sophisticated, is easier to fathom than gut forecasting, if for no other reason than it can be explained in terms of tools to use that have cause-effect relationships.

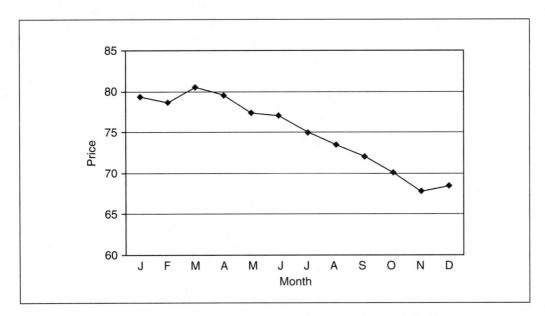

Figure 3–15   Seasonal price pattern for cattle in the U.S. (2001)

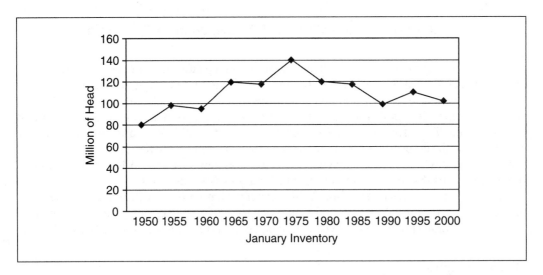

Figure 3–16   Cyclic pattern for cattle in the U.S. (1950–2000)

Econometrics is the study of quantifying economic relationships. A combination of mathematics and statistics forms the core of the tools used. An economic relationship is expressed in a mathematical format and then a statistical tool is used to estimate the values of the mathematical model. The statistical tool most often used is multiple regression analysis. The process involves the following:

1. Determine the economic relationship. Suppose the relationship between beef and pork is needed. The economic relationship is stated as: What does the demand for beef do when the price of pork changes? If beef and pork are substitutes, then as the price of

pork goes up, the quantity demanded of pork will decrease (a movement along the demand curve for pork). Consumers will consume less pork, and because they consider beef to be a substitute for pork, they will shift some of their consumption to beef. This will cause the demand curve for beef to shift right, or increase. This is first-round effect. But, as the demand for beef increases, the price of beef will increase, assuming other factors are held constant. The rising price of beef might then cause a series of second-round effects with other substitutes (chicken or fish perhaps) and so on. At this point, forecasters must decide how complicated they want to make the economic model. Most forecasters stop with only the first-round effects. Other forecasters will simply ignore the potential effects of pork prices on beef prices because they consider them to be small and not worth the effort to quantify.

2. What is the mathematical expression of the economic model? If a forecaster wanted to determine the price of beef, she might start by creating a proposed mathematical expression as

$$P_b = f(Q_b)$$

where

P_b = price of beef

Q_b = quantity of beef

f = Some as-yet undetermined mathematical functional form

This is a simple model that may or may not correctly specify the economic relationship between price of beef and the quantity of beef, but could be a good predictor of beef prices. Once the variables have been specified (P_b and Q_b), the mathematical equation must be specified, that is, what form does f take? Is it a simple linear relationship? Logarithmic? Quadratic? If it were assumed to be linear, then it would be expressed as

$$P_b = a + b_1 Q_b$$

where

a = intercept

b_1 = slope of Q_b relative to P_b

3. Determine what data to use and the time frame of analysis. How long of a time period is needed? Will the data used be daily, weekly, or monthly? Will the data be for a specific market or aggregated over several markets? If daily or weekly data are used, what allowance is made for incomplete data series (i.e., holidays when markets are not open)? If the data is aggregated, which markets are included and which excluded?

Econometrics, to be sure, is many times more complicated than this simple explanation. But no matter how complicated the process, three major areas are at the crux of each analysis: (1) What is the economic relationship? (2) How can it be expressed mathematically? and (3) What data are available and can be used? Some forecasters use very simple models while others use extremely complex ones. It is important to notice that even a simple model that expresses the price of beef as a function of the quantity of beef still requires a significant investment of time in development of the model, specifying the form, and collecting the data. And if the results are poor, refinements will have to be made in the model, mathematical form, or the data used. Needless to say, most econometric work is performed at universities or large consulting firms that have the expertise to put all of the elements together. Little wonder that most traders shun econometric analysis and concentrate on simple cause-effect relationships that they can more easily apply themselves without extensive analysis or outsourcing of the statistical work.

Technical Price Analysis

Technical analysis is based on the belief that where prices have been in the past can be used as a guide for the future direction. Technical analysis does not directly dispute fundamental analysis but plainly believes that all of the information is embedded in the price movement and that it is impossible to fully determine all of the factors that influence price. *A technical analyst would say that fundamental analysis determines the reasons why market prices move and technical analysis studies the effects.*

Two major types of technical analysis are used by the majority of technical traders—charting and mathematical modeling. Charting entails the physical recording of price in some visual form and then studying the past price profile. Mathematical modeling is sometimes called "form fitting" because it is the process of trying to find a mathematical formula that will mimic the past price movement.

Charting Analysis

The major idea with charting is to visualize price information. One of the first to use charts to determine price direction was Charles Dow in the late 1800s. Charles Dow used price charts to develop what was later dubbed the Dow Theory. The theory states that a major bull market (a rising market) will continue as long as the intermediate highs and lows continue higher as depicted in Figure 3–17. Likewise, a major bear market (a declining market) will continue as long as the intermediate highs and lows continue lower. Mr. Dow's theory, as well as many other tenets he proposed and used to analyze stock prices, is still widely used by technical analysts.

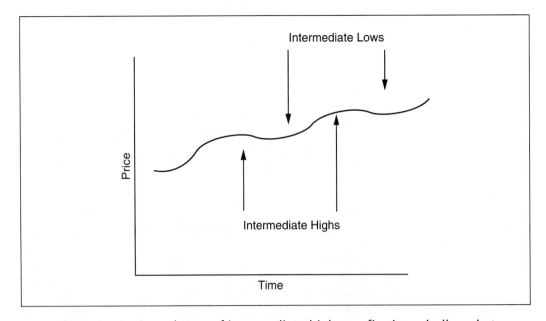

Figure 3–17   Dow theory of intermediate highs confirming a bull market

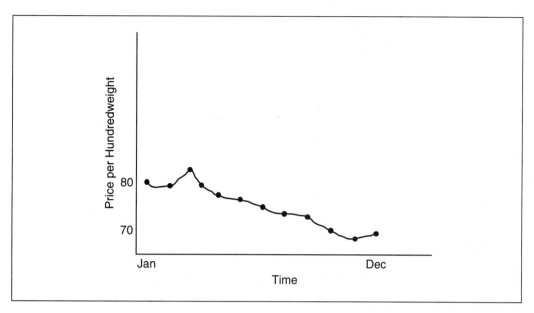

Figure 3–18   Single price chart for cattle (2001)

Chart Types

Single Price Charts   A single price chart merely records a price for a designated time period. This price can be the average price for the time period, such as a day or week, or it can be any other price. Often the price used is the closing price for daily price charts, and these charts are referred to as closing price charts. There is no one right price to use in single price charts, as long as the same price is used consistently throughout the time period of the chart. Figure 3–18 demonstrates a single price chart.

Single price charts are the most common charts used when mathematical tools are used to analyze prices. They are also the most common charts used when mechanical trading systems are created using a computer.

Bar Charts   Single price charts contain limited pricing data; a serious chartist prefers more information. Therefore, for serious chartists, the most common form of charting is a bar chart. A **bar chart** records the high, low, and settle (or close) as portrayed in Figure 3–19. The most widely used bar chart is a daily chart that records the high for the day, the low, and the settle (a weighted average of the last few minutes of the closing prices). Inter-day bar charts are also used by some traders whereby they record the high, low, and close for a designated time period such as from 10 A.M. to 11 A.M. Regardless of the time frame used, a bar chart will show the range of price movement on the vertical axis and time on the horizontal axis.

Point and Figure Charts   The second most popular way to visualize prices is a **point and figure chart.** The point and figure chart records the price movement regardless of time. The vertical axis records the price level and the horizontal axis *de facto* records time, but does not do so for a specific time. Point and figure charts use "Os" to record falling prices and "Xs" to record rising prices as illustrated in Figure 3–20. Since the horizontal axis is not a specific date, point and figure charts reflect only price movements regardless of time while a bar chart reflects price movements as a function of time.

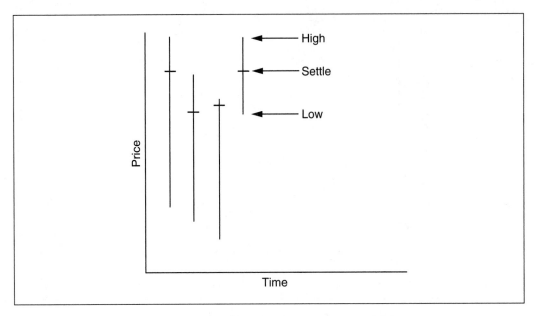

Figure 3–19   Bar chart

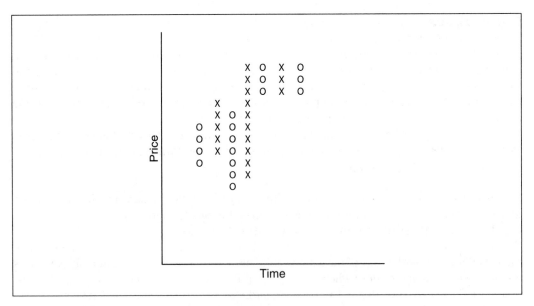

Figure 3–20   Point and figure chart

Point and figure charts must denote a box size to represent a price movement, such as 5 cents. If a price move occurs that is smaller than the box size, the movement is not recorded. Additionally, a reversal criterion (a change in direction) must be stated such as "two box sizes." For a price direction to be reversed and recorded, the reversal criteria must be met. Otherwise, the price movement will not be recorded. For a two-box reversal with

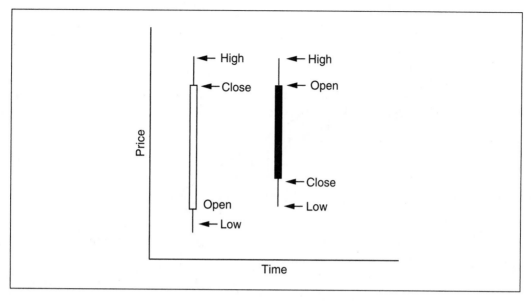

Figure 3–21   Candlestick chart

each box representing 5 cents, a price reversal must be at least 10 cents in the opposite direction to be recorded as a directional change.

Candlesticks   A minor charting tool called **candlesticks** has emerged during the last few years. A candlestick will record the basic information of a bar chart and also add the opening price and the closing price. A candlestick will show the close and open as a two-dimensional body and the high and low as lines called shadows, shown in Figure 3–21. If the close is higher than the open, the body will be white (green) and if the close is lower than the open the body will be black (red). Each candlestick represents a specific trading day, just like bar charts.

The purpose of each of the charting tools is merely to express visually what a price movement looks like in order to determine two major things: (1) how long a trend (direction) will continue, and (2) when a trend will reverse. Over the years an almost uncountable number of tools have emerged (enough to fill hundreds of books) to determine these two major price movements.

Trends and Turning Points

How long will a bull market continue? Is this bear market a short-term trend? These are questions that chartists attempt to answer. The tool most often used is a trend line as indicated in Figure 3–22. Trend lines have some of the concept of the Dow Theory in them. A bull trend will continue as long as the intermediate lows keep getting higher and thus do not break the trend line and may ultimately add another point to the trend line.

Other tools are also used to indicate that a trend will continue such as flags and pennant formations, triangles, and runaway gaps. All three of these concepts indicate to a chartist that a trend is approximately half way completed. Figure 3–23 shows a flag formation.

A reversal in the trend is beaconed when the trend line is broken by a price movement as displayed in Figure 3–24. Other popular reversing formations include exhaustion gaps and double tops and bottoms. Figure 3–25 shows an exhaustion gap.

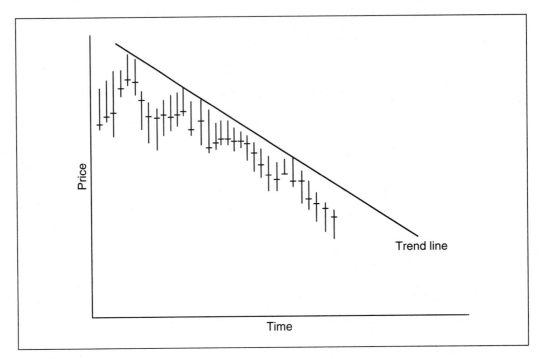

Figure 3–22 Bear trend with trend line

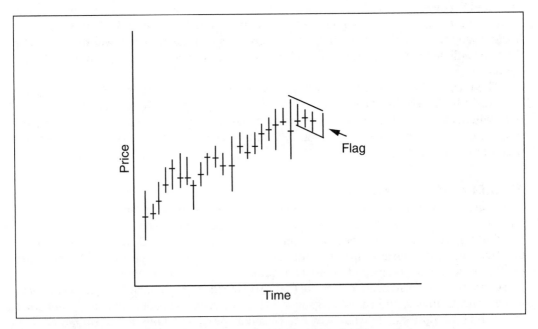

Figure 3–23 Flag formation in a bull trend

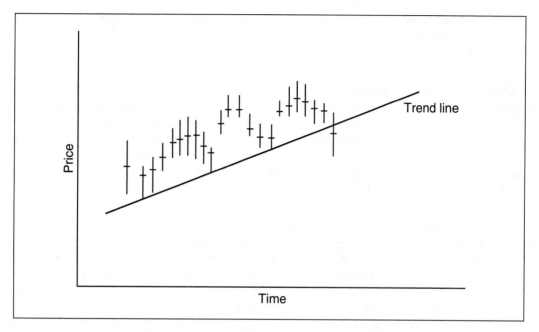

Figure 3–24   Bull trend line with violation

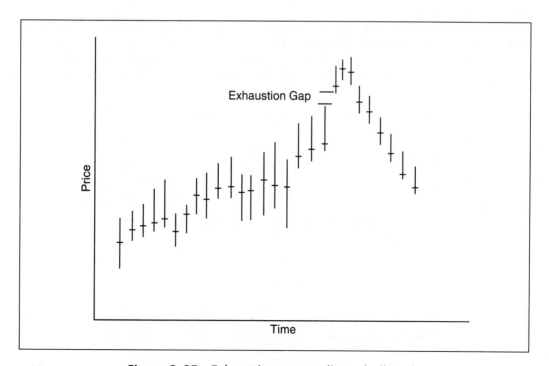

Figure 3–25   Exhaustion gap ending a bull trade

The list of possible formations is almost limitless and no attempt is made in this chapter to cover more than a few of the most widely used tools so that the reader can get an idea of some of the tools and how they are used by a chartist.

Mathematical Modeling

Because charting analysts most often attempt to answer the questions of trend duration and turning points in price moves, another form of technical analyst has emerged, especially as computer speed and capacity has increased in the last few years. Mathematical modelers or mechanical analysts have become very popular as they attempt to forecast prices based on previous price information. Mechanical analysis falls into three major categories: **curve fitting, moving averages,** and **oscillators.**

Curve Fitting

Curve fitting is the general expression used to imply that for a given set of past price movements, an equation will be selected that fits the data the best. The selection of an equation can be done visually or by statistical estimates, usually with multiple regression. Figure 3–26 shows a simple linear model's forecast of the next price. The example is a simple linear regression model that estimates the best fit of the actual data to an estimated line.

More complex mathematical models have evolved over the years. In fact, in the last several years advances in chaos and complexity theory have created a following of curve fitters who use the relatively new mathematical tools to fit economic data in an attempt to be able to forecast market price.

Moving Averages

By far the most used mathematical tool for price forecasting is the rather simple concept known as a moving average. The idea is to calculate at least two averages of past prices, one a short-term average such as the last three days and a longer average such as ten days. When the two averages cross each other, a change in the trend is indicated. Three averages are

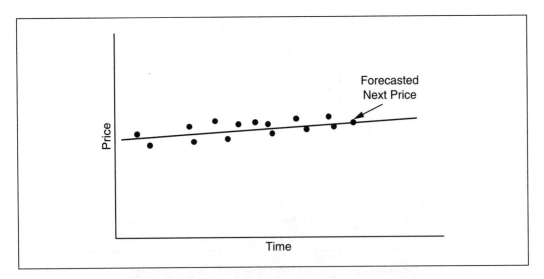

Figure 3–26   Best fit model forecast of next price

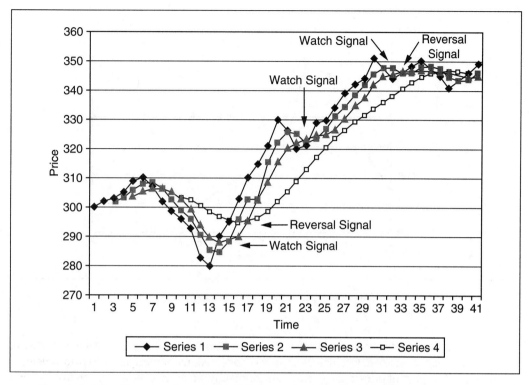

Figure 3–27   Three-, nine-, and eighteen-day moving average model where
Series 1 is the actual price, Series 2, the 3-day average, Series 3,
the 9-day average, and Series 4, the 18-day average.

quite popular whereby the short and intermediate average crossing implies a watch signal for a possible trend change; when the intermediate average crosses the longer average, a trend is deemed over. The very popular 3-, 9-, and 18-day averages are shown in Figure 3–27.

Moving averages are extremely popular because they are very easy to calculate and use with or without a computer. Some analysts use weights to change the averages according to some arbitrary criteria in an attempt to add accuracy.

Oscillators

An oscillator is an elementary arithmetic expression used to measure the rate of change of prices. The simplest one is called a momentum chart. The formula is

$$M = P - P_x$$

where

M = momentum

P = current price

P_x = price at time period x days ago

If M is negative then the current price is below the price at some previous time and if it is positive then the current price is above the previous price. These plus and minus differences are then plotted around a zero line, as illustrated in Figure 3–28.

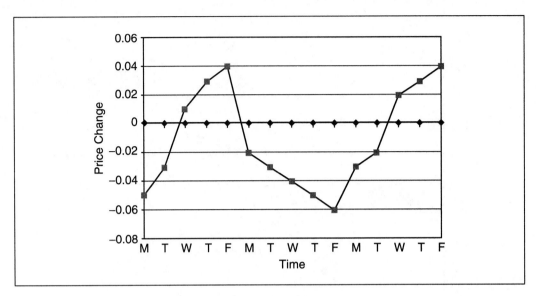

Figure 3.28   Momentum oscillator

Other popular oscillators are relative strength and stochastic indexes. However, the list of oscillators is long and varied. Oscillators are uncomplicated mathematical expressions used to measure some aspect of price change and thus, hopefully, be more useful to price forecasters.

A Final Word

At some point, all of us will make a price forecast. Sometimes it is as simple as whether or not to fill up the gas tank on the car now or wait a few days when we forecast (hope) that the price will be lower. It is critically important for businesspeople to seriously think about what they believe about price forecasting. If they believe it is impossible to forecast prices with any accuracy, then how they construct their marketing plans and capital outlay budgets will reflect their attitude toward the risk of price change. Conversely, if they believe that prices can be forecast with some accuracy, alternative price risk management strategies will need to be developed. How people feel about price forecasting is directly related to how they will handle the risk of price change.

CHAPTER 3—QUESTIONS

1. If a trader believes in the Efficient Market Hypothesis, would he ever use Fundamental Price Analysis?

2. How can both technical and fundamental analysis be used by a trader?

3. What is the major difference between bar charts and point and figure charts?

4. What would be a good example of a demand driven product? Why?

5. What kind of product would have a very inelastic demand curve? Why?

Markets, Exchanges, and Regulation

This chapter covers the basic economic factors that influence futures markets and a brief listing of the major agricultural futures exchanges and the regulating groups associated with the futures industry.

OVERVIEW

Modern day futures markets evolved from early cash markets as the risks of price changes associated with storage, transportation, and processing of agricultural commodities were securitized in various futures contracts. To assure the financial integrity of the new contracts, exchanges evolved as mutual organizations. A mutual is an organization whose members share equally in the profits and losses or in the running of the business if it is a nonprofit entity. All of the major North American futures exchanges were originally set up as mutual organizations and only in the early 2000s did they start the process of demutualization and the conversion to for-profit publicly traded corporations.

Markets

Central places to buy and sell items have existed longer than recorded history. Archaeological evidence of trading centers pre-dates recorded history by several thousand years. Market activities are some of the first items recorded as humans embarked on written language. Most modern markets today are little changed from those of several thousand years ago. Buyers and sellers come together in one physical location. Buyers can inspect the seller's merchandise and price is individually negotiated. That is the essence of all markets; buyers have to be satisfied that the seller's product meets their needs and the price is acceptable to both parties.

When buyers can physically inspect the product and use their own skill and judgment as to the condition and quality of the product, the general market rule has been in common law to be *caveat emptor* (Latin for "let the buyer beware"). Unethical sellers throughout history have preyed on this concept by various methods to fool or trick the buyer (watered wine, sand in grain, horse meat instead of beef just to name a few). The acceptance of *caveat emptor* into common law throughout history is hinged on the belief that buyers and sellers are of equal power and the buyer was free to inspect the item and thus had control over the transaction and was the responsible trader.

As economies became more complex, so did marketplaces. Buyers and sellers alike needed more than just a spot (cash) transaction that involved physical movement of the product to a central trading location and then a time-consuming inspection and price negotiation. Gradually, time (delivery) became part of the negotiation as well as alternative payment methods. Two general types of time contracts emerged that are still in use today: **to arrive contracts** and **forward contracts.** To arrive contracts call for delivery at some point in the future with title passing immediately from seller to buyer. Forward contracts call for delivery at some point in the future but title passes only upon delivery. Timed delivery created a new market participant—third-party handlers. These third-party handlers would perform the function of storing, delivering, and sorting of products (merchants, warehouses, and commercial transportation). Alternative forms of payment evolved from simple cash transactions to negotiable orders of withdrawals, checks, letters of credit, and purchase orders. These alternative forms of payment between buyers and sellers necessitated the development of other areas of professionals such as record keepers (bookkeepers and accountants), bankers, money changers (foreign exchange), and mediators (attorneys).

As time and payment choices were added to central markets, another axiom emerged—*caveat venditor* (Latin for "let the seller beware"). Under this rule, the seller is the responsible party and has to act accordingly. With both caveat principles in the minds of traders, a need developed whereby both buyers' and sellers' rights were honored in a more formal structure that also accommodated more complex trading terms.

Such markets developed in the mid-1800s entailing agricultural commodities throughout the United States, Canada, and Europe. Buyers, sellers, and the commodity would still be centrally located but certain rules of conduct and financial assurances were imposed by the central market to provide some measure of relief between buyer and seller on the risk of default. Livestock markets arose near water or rail routes so that buyers could move large volumes of animals to slaughter and large numbers of sellers would provide the pool of animals. Live auctions developed at these markets as an efficient way for many buyers to have access to all of the seller's products and sellers could be assured of more than one buyer having a chance to bid on purchase price.

By contrast, grain markets unfolded along a different path. Sellers would privately negotiate with a third-party handler (often a warehouse). In turn, buyers would do likewise with the third-party handler. Sellers (grain producers) needed to get rid of the product after harvest, and warehouses (later called elevators) would buy the grain and then perform the value-added function of storing the product. Buyers could buy from the warehouse throughout the year and avoid having to store the product themselves.

Economic Theory and Markets

Modern microeconomic theory classifies markets into three major forms: **perfectly competitive, oligopoly,** and **monopoly.** An understanding of each model will help understand the development of markets and how they have evolved to today.

Perfectly Competitive Markets

For markets to behave in a perfectly competitive way they have to have the following:

1. Many buyers and sellers with no one being large enough to have any market power over the others

2. A homogenous product

3. No barriers to entry or exit

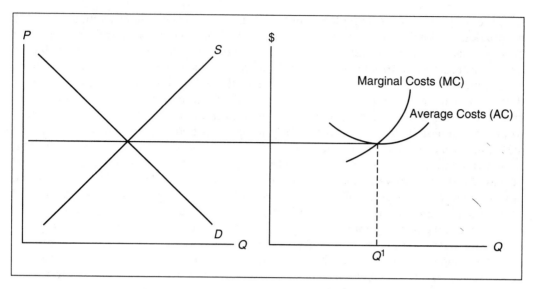

Figure 4–1   Perfect market model with no excess economic profit

The results of such a market are that the price in the short run oscillates around the cost of production and in the long run no excess economic profits exist.

In this market, traders can enter the market or exit at will. There are no legal, economic, or intellectual barriers. No seller can differential their product from another, that is, everyone is selling the same thing. There are many buyers to offer bids and many sellers all with the same product to offer and all of the buyers and sellers are approximately equal in size and power. Obviously, no known markets completely adhere to these assumptions. Most economists relax the assumptions and reclassify the perfectly competitive market as *workably competitive*. Workably competitive markets don't exactly meet the requirements, but behave or have the results expected from a perfectly competitive market.

In such a market, the interaction of the supply and demand curves yields the market price that, in turn, becomes the marginal revenue to each producer, as shown in Figure 4–1. Since no producer is big enough to affect price, each unit that the producer sells brings in the same revenue, thus the reason the market price is the same as the individual producer's marginal revenue curve. The individual's cost structure shows in Figure 4–1 that there is no profit. If something occurs to affect market demand such that the market price increases as shown in Figure 4–2, then the individual producer will have short-term profits. However, since there are no barriers to entry or exit, then the short-term profit will attract other producers (shifting the market supply curve to the right) and thus drive the market price back to where there is no profit for an individual producer as revealed in Figure 4–3.

The situation is the same if there is a loss. Producers will leave the industry and price will increase. Thus the competitive model yields the potential for short-term profits and losses for individual firms, but in the long run market price tends to seek the level associated with the total cost of production, thus no excess economic profits.

Not Perfectly Competitive Markets

An oligopoly is two or more firms producing a product that may or may not be homogenous and there are some form of barriers to entry or exit in the marketplace. A monopoly is a single producer with significant barriers to entry. Both types of markets are considered

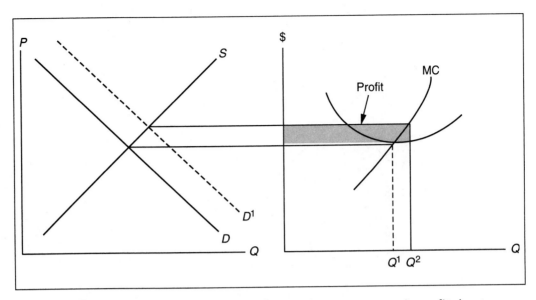

Figure 4–2   Perfect market model with short-term economic profit due to
a change in demand

noncompetitive because some market power exists. If oligopolies differentiate their products and do not enter into price wars with each other, they can sustain prices such that they have excess economic profits. Obviously a monopoly can do likewise. If a market sustains short-run or long-run excess economic profits, one or more of the assumptions of a perfectly competitive market have been violated. Oligopolies and monopolies refer to the production side of a marketplace, that is, the sellers of the product. A single buyer is called a **monopsony** and two or more buyers are referred to as an **oligopsony.**

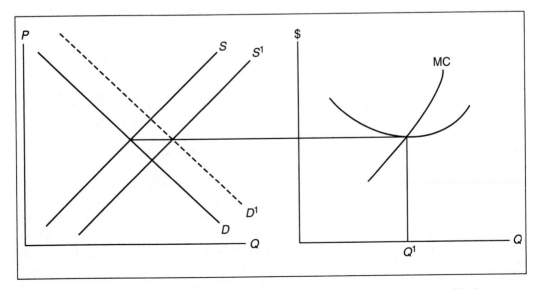

Figure 4–3   Perfect market model with no short-run economic profit due
to additional market participants

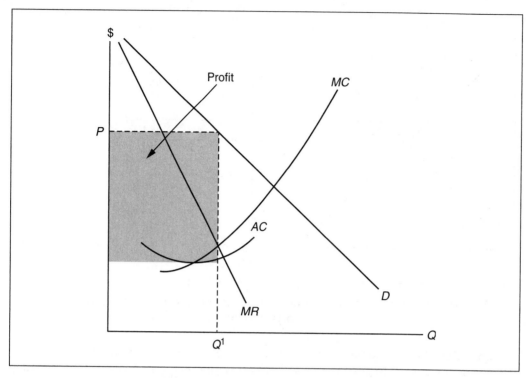

Figure 4–4   Monopoly market model showing excess economic profit

Let's consider the situation of a monopoly first and then treat oligopolies as a special case of the monopoly model. Figure 4–4 shows the cost structure for a single firm and that they will price the product at the level that maximizes their profit (where marginal cost equals marginal revenue). Since they are the only producing firm, they can maintain that price and profit level. Consequently, most governments will endeavor to put limits on what monopolies, such as utilities, can earn so they cannot behave as a monopoly.

The monopoly model is useful in agricultural situations because sometimes a local business can behave in a manner not unlike what the monopoly model says will happen. For example, if there were only one feed processor in an area they could set their price at a level that would produce excess economic profits, even for the long run, if there were enough barriers to entry to keep other competition out. Rural areas do have businesses that behave in a fashion similar to monopolies, which over the years has caused farmers and ranchers to form cooperatives as a way to counter the monopoly practices of businesses.

Consider now the case of oligopolies' (more than one but less than many) producers. They try to differentiate their products such that consumers believe there are no substitutes, thus they can have some measure of monopoly profits. To the extent that consumers don't feel the products of two producers are different, then the two firms will compete with each other. If they do so on price, they might behave more like the competitive model where profits are bid away and over time no excess economic profits exist (the current situation with the airline industry). They might try to work together and avoid a price war and then divide the excess economic profit based on market share; however, that is illegal and oligopolies are constantly being monitored both at the federal as well as the state level for such actions of collusion.

The point of studying the economic models is that the only way to achieve long-run economic profit is to have some form of noncompetitive market activity. Most firms try to do this by differentiating their product (Coke versus Pepsi) or driving their cost structure so low that the market price is above their costs such that they can have excess economic profits beyond the short run (Southwest Airlines and Wal-Mart).

Using the Market Concepts and the Role of the Speculator

The various market forms and the economic assumptions are helpful to understand market behavior and are critical for market participants called **arbitragers.** An arbitrager is a trader who will attempt to take advantage of market imperfections in order to capture a profit. Arbitragers are also known as *profit takers.* Arbitragers are **speculators** who have a very sophisticated knowledge of markets and how they function and provide a vital role of the efficiency of market behavior.

Arbitragers will generally avoid markets that are characterized as either a monopoly or monopsony simply because one buyer or seller is *de facto* the market. One buyer or seller exists because of significant barriers to entry that preclude any other market participants. However, other than monopolistic markets, arbitragers can and do participate to the extent that they believe they can earn a profit for their knowledge and actions. Arbitragers can be classified in two major ways: (1) market relationships and (2) market positions. Relationship arbitragers (also known as **spreaders**) are looking for abnormal patterns in market associations such as time, place, and form. Position arbitragers take positions in the market believing that the market will move in favor of the position. Let us look at examples of both types of arbitragers.

Market Relationships

Temporal

The difference between two markets that are separated by time and where the product is storable, if the markets are workably competitive, should be the cost of carry between the two time periods. For example, the price of a bushel of corn today in the cash market should be lower than the price of corn to be delivered in one month to the same location by the cost of storage (carry) for one month. If the cost of carry for one month was 5 cents per bushel, then the relationship between cash spot price today and one month later should be 5 cents per bushel. Consider how market forces maintain that relationship. If spot cash corn was trading at $3 per bushel today and the one-month delivery for corn to the same location was trading at $3.08 per bushel, arbitragers would enter the market by buying cash corn today and simultaneously selling it for delivery one month later for $3.08 per bushel. They would store the corn for the month incurring a cost of 5 cents and earning a tidy 3 cents per bushel profit. This action will be done by several traders. The markets will likely react as follows: the extra buying pressure on the cash market will bid the local cash price up and the simultaneous selling pressure for one-month future delivery will bid the price down, and the 3-cent profit will be bid away by the actions of the arbitragers. If the markets did not respond this way, arbitragers would look at alternative market models such as monopoly power to determine if, in fact, the markets were competitive.

Spreaders use this concept all the time with storable commodities on futures contracts. An arbitrage spreader will determine what is "normal" as a price difference between two or

Table 4–1   Normal Temporal Spread

Buy March corn futures @ $3.00/bushel
Sell May corn futures @ $3.14/bushel

Spread of 14 cents/bushel

Later…

Sell March corn futures @ $3.02/bushel
Buy May corn futures @ $3.12/bushel

Spread of 10 cents/bushel

Profit of two cents per bushel on the March leg (buy @ $3.00, sell @ $3.02)
Profit of two cents per bushel on the May leg (sell @ $3.14, buy @ $3.12)

Total Profit = 4 cents/bushel

more futures market prices (the spread) and then watch for times when the spread is abnormal. If, for example, a spreader determined that the normal spread between March corn futures price and May corn futures price was 10 cents per bushel, they would watch for times when the spread was either less or more. If they observed that March corn futures was trading at $3.00 per bushel and May corn futures was trading at $3.14 per bushel, what should they do? They have determined that the normal spread is 10 cents (i.e., the cost of carry for two months) and then the spread suddenly widens to 14 cents. The theory says that the market will go back to the 10-cent spread. A spreader will simultaneously buy the March corn futures at $3.00 per bushel and sell the May corn futures at $3.14 per bushel. They have "put on a spread" by buying the nearby futures contract and selling the faraway contract. Then they wait. Their actions combined with those of other spreaders will most likely force the markets to adjust. There is buy pressure on the March contract, which will increase the price, and selling pressure on the May contract, which will lower the price and thus move the spread back to 10 cents. Table 4–1 shows this action. All that matters is that the spread goes back to the normal spread and the spreader will make as a profit the difference between the initial spread and the abnormal spread. The general rule is that when the spread is larger than the normal spread, the arbitrager will "put on a spread" which means that they simultaneously buy the nearby futures contract and sell the faraway futures contract.

If the spread becomes abnormally small, a spreader will do just the opposite, as revealed in Table 4–2. When the temporal spread becomes smaller than normal, the arbitrager will "put on reverse spread" whereby he or she simultaneously sells the nearby futures contract and buys the faraway futures contract. The selling of the nearby futures contract should cause the price to decrease and the buying pressure on the faraway should increase that price such that the spread increases.

Notice that in either the case of the normal spread or the reverse spread the profit earned is the difference between the normal spread and the abnormal amount. This action by spreaders helps keep the futures markets tied together by the cost of carry market model. The spread model is stronger than the reverse spread. If a spread is wider than normal and the markets don't go back to the normal spread, the spreader can take delivery of the nearby commodity, store the product until the faraway date, deliver at the later date, and capture the difference. However, if the markets are abnormal and the spreader puts on a reverse

Table 4–2   Reverse Temporal Spread

Normal spread March-May = 10 cents

Sell March corn futures @ $3.00/bushel
Buy May corn futures @ $3.07/bushel

Spread of 7 cents/bushel

Later...

Buy March corn futures @ $2.99/bushel
Sell May corn futures @ $3.09/bushel

Spread of 10 cents/bushel

Profit of 1 cent per bushel on the March leg (sell @ $3.00, buy @ $2.99)
Profit of 2 cents per bushel on the May leg (buy @ $3.07, sell @ $3.09)

Total Profit = 3 cents/bushel

spread, the option of delivery is not available since the spreader cannot sell physical commodities that he does not have and buy later. In other words, the trader cannot have negative inventory in the storeroom. Accordingly, spreads are less risky than reverse spreads.

Spatial

The difference between two markets that are separated by space should be the cost of transportation. Arbitragers would trade the differences if they were aberrant from the cost of transportation, just as they would for time differences. If the price of a bushel of corn was $3 in one location and $3.20 in another, and the cost of transportation between the two locations was only 15 cents per bushel, arbitragers would buy in the cheaper location and pay to transport to the other location and earn 5 cents profit for their actions. These actions should drive up the price in the cheaper location and down in the dearer location so that the difference represents the cost of moving the corn between the two locations.

Unlike temporal spreaders with futures contracts, no major special agricultural futures spreads exist as there are no agricultural products that are alike traded on two different exchanges separated by place.

Form

The difference between the price of the raw product and the finished product(s) should be the cost of form change (manufacturing costs). The price difference between raw soybeans and the meal and oil that result from crushing the beans should be the cost of processing. Arbitragers would watch the relationship and trade when the difference was less or more than the cost to transform raw products into finished products.

Form spreads exist in the energy field as crude oil is refined (cracked) into gasoline, heating oil, and other products. In fact the New York Mercantile Exchange has a cracking futures contract based on the cost of transforming (cracking) crude oil into its component parts.

Live cattle futures represent the final product while feeder cattle are an input into the process as well as the feed necessary to produce the fat cattle such as corn and soybeans. It is conceivable that a cattle crush or crack spread exists (a feeding spread), but thus far it has yet to be quantified in such a way as to be valuable to spread arbitragers (or if they have found it they are certainly keeping their mouths shut). Likewise it is conceivable that a similar relationship exists between raw milk prices and its component parts of butter and cheese.

What is necessary for successful spreading from differences is a futures contract on the raw product(s) and finished product(s) so spreaders can calculate what is normal as a processing or transforming cost and the prices of raw and finished products. Soybean crushing by spread arbitragers is very popular because futures contracts exist on the raw product (soybeans) and finished products (meal and oil). Soybean crush hedges are discussed fully in Chapter 5, but the concept is the same regardless of whether the action is a hedge or spread.

Spread arbitragers will spread relationships between two or more cash markets and likewise for two or more futures markets, and will also trade on differences between a cash market and a futures market price. These actions keep the cash and futures markets tied together by the three major relationships—time, space, and form.

Position Traders

Position arbitragers will take a market position if they believe the market is undervalued or overvalued other than by the three major relationships of time, space, and form. Arbitragers might sell corn for future delivery right before a major governmental report because they believe the report will be bearish. If the report is bearish, then they will have sold the corn before the report and can now buy the corn at a cheaper price to deliver against the contract. Position arbitragers will buy one type of cattle in one market because they know that type of cattle is always sold at a discount in that market, and sell the cattle into another market that pays a premium.

Arbitragers provide two major market activities. They provide liquidity because they are constantly monitoring markets and trading. They are also the glue for economic activity. Arbitragers are ready to exploit differences in time, space, form, discriminations, information, local monopolies, or any abnormal market situation to earn a small profit for their efforts and knowledge. Thus successful arbitragers are very sophisticated in their understanding of market forces and market participants.

Because arbitragers are ready, willing, and able to trade off differences between cash markets and futures markets, they force the two markets to be tied together in a derivative relationship. A futures contract derives it value from the underlying cash market, thus the reason early cash markets such as the Chicago Board of Trade evolved other forms of derivative contracts such as futures and options contracts.

The Exchanges

Exchange traded derivatives have gone through major changes followed by long periods of almost no change. The first major movement involved the actual formation of organized **exchanges** where the rules of trade could be developed and a central place existed to physically trade commodities. The first to form a central trading place complete with rules of trade and conduct was the Chicago Board of Trade (CBOT). Once the CBOT got started they created the first futures contracts, probably vintage civil war times (mid-1860s). The exact date is uncertain because the CBOT lost all of its records in the great fire of 1871.

Other exchanges developed during the period between the Civil War and the beginning of the twentieth century. Numerous commodities were traded and indeed, whole exchanges were created around a few products such as the Chicago Butter and Egg Board (later to be renamed the Chicago Mercantile Exchange). Exchanges were formed in Kansas City, Minneapolis, and New York, but they did not do anything new with futures contracts other than make futures contracts more regionally focused.

The only major change that occurred with futures exchanges between the Civil War (1860s) and World War II (1940s) was the formation of a separate clearinghouse. In 1926 the CBOT created a clearinghouse that was a separate unit from the trading exchange. Other exchanges followed, and in an interesting twist in 2004 the CBOT moved all of its clearing functions to the Chicago Mercantile Exchange (CME). The clearinghouse concept was a major shift as it split the financial risk of default on the futures contracts from the function of trading the contracts.

Futures markets, once developed in the mid-nineteenth century, remained virtually unchanged until the mid-1960s when the CME developed the first futures contracts on live animals—hogs and cattle. The CME followed in 1972 with the creation of a new division within the CME called the International Monetary Market (IMM) to trade the newly floating foreign currencies. Within a decade the CME had innovated new futures contracts on non-storable commodities and entered the world of financial derivatives. The CME introduced the first cash settled contract (Eurodollar contract) and the first index contract (S & P Stock Index). Other exchanges quickly followed suit with an almost infinite stream of index and financial products. The flood gates were opened by the CME; essentially anything that is traded in some way can have a derivative contract attached. The period between the early 1980s and the beginning of the twenty-first century is remarkably similar to the same time period one hundred years earlier—lots of new futures contracts on many different types of products, but no real structural or financial changes.

Surprisingly, during the period when the CME was developing new innovative futures contracts, none of the exchanges seriously looked at options until a rough group of traders in the early 1970s found a loophole in the Commodity Exchange Act of 1936 that banned options to open option trading on commodities. The Commodity Futures Trading Commission (CFTC) moved to close the loophole and ban all commodity options once again by the mid-1970s. The CFTC allowed a new pilot program on options on commodities in the early 1980s and exchange traded options quickly moved into the mainstream of derivative use.

Computerized trading is embraced throughout the world, yet it remains a stumbling block with U.S. exchanges. The U.S. exchanges have steadfastly refused to accept computerized trading as a substitute for open outcry auctions. The exchanges have tried to use handheld devices as a way to computerize while still keeping the physical trader in the pit. These devices met with dismal failure because they failed to use the power of computers and instead forced a new hard-to-use technology on traders whose biggest asset was the speed with which they could execute a physical open outcry trade. The uneasy truce that has evolved in the United States is a kind of dual system that allows for off-hour computerized trading and an active open outcry auction in the pits—neither a feast nor a foul bargain.

The latest change to occur with U.S. exchanges before 2000 had been the **demutualization** of the two largest exchanges. In 2000 the CME voted to become a for-profit, publicly traded stock company. In 2002 the process was complete, and the public can now own shares of the CME. The CBOT voted in 2004 to demutualize and the process was complete in 2005. All of the other U.S. exchanges remain **mutual** not-for-profit companies. All of the exchanges have excellent Web sites, many with downloadable data and more detailed information on futures and options contracts.

Chicago Board of Trade

The CBOT is the granddaddy of U.S. exchanges (often referred to as the "Board"). It was the first to form in 1848 as a place for traders to assemble and conduct business under mutually accepted guidelines for the professional conduct of business. The forerunner to futures contracts were called "to arrive" contracts. Traders negotiated quantity, quality, and price and then set a time for the product to be delivered ("to arrive"). Since the traders were all assembled in one location, these early contracts started to take on standard features that would allow for easier retrading, such as quantity and quality. Once the major features of the contracts were standardized, all that remained to be negotiated was price between traders. Exactly when these standardized "to arrive" contracts slipped into the form known today as futures contracts remains obscure because the early records of the exchange were destroyed in the Great Chicago Fire of 1871 (over a third of the city was destroyed including 18,000 buildings); however, the CBOT believes that sometime during 1865 the first futures contract began trading. The exchange trades a wide range of futures and options contracts but is generally known for its derivatives on grains and soybeans. The CBOT trades fertilizer contracts and in 2005 it added an ethanol contract. The Internet address for the CBOT is <http://www.cbot.com>.

Chicago Mercantile Exchange

The "Merc" started as the Chicago Butter and Egg Board in 1898 and officially changed to its present name in 1919. The CME has long been an innovator in the derivatives world. They started the first nonstorable commodity futures contract, live cattle, in 1964 and in 1966 followed up with a live hog contract. They were the first exchange to move into financial derivatives in 1972 with currency futures contracts. The CME built the first global **electronic trading** platform, GLOBEX, in 1992. U.S. exchanges have always been mutual nonprofit associations, but the "Merc" stopped that in 2002 by becoming the first exchange to demutualize and become a for-profit, publicly traded company. The CME became the world's largest derivative exchange in 2004 by trading a long and diverse list of products, but it is primarily known in the agricultural arena as the livestock exchange. The Internet address for the CME is <http://www.cme.com>.

New York Mercantile Exchange

The NYME (pronounced "nyemex") is one of the nation's oldest exchanges. It was founded in 1872 to trade butter, eggs, and cheese. It has long since abandoned those products and emerged in the last half of the twentieth century as the metals market trading gold, silver, copper, platinum, and palladium. The NYME made a major jump in the late 1970s by offering energy derivatives—crude oil and gasoline futures contracts. Many additional energy derivatives as well as some financial indexes have been added, and today the NYME is the premier energy derivative exchange in the world. The Internet address for the NYME is <http://www.nymex.com>.

New York Board of Trade

The NYBT came into being in 1998 via the merger of two very old exchanges—the Coffee, Sugar and Cocoa Exchange (founded in 1882) and the New York Cotton Exchange (founded in 1870). Today the NYBT is a major world derivatives player because it offers derivatives in cocoa, coffee, orange juice, sugar, and milk. It also has numerous financial derivative products. The Internet address for the NYBT is <http://www.nybot.com>.

Minneapolis Grain Exchange

This exchange has been the major market for hard red spring wheat derivatives since its founding in 1881. The MGE innovated the first agricultural commodity indexes in the early part of the twenty-first century and now trades indexes on corn, soybeans, hard red spring wheat, hard red winter wheat, and soft red winter wheat. These indexes remove specific delivery locations from the standardized component of the contract and thus are excellent tools to follow the general market price of these commodities. These indexes provide a "fixing" market for cash settlements in other markets (necessary for successful swap contracts). The Internet address for the MGE is <http://www.mgex.com>.

Kansas City Board of Trade

The nation's second oldest exchange was founded in 1856. The KCBT's major derivative product is hard red winter wheat, but in 1982 it launched the first financial derivative index futures contract, the Value Line. The Internet address for the KCBT is <http://www.kcbt.com>.

Winnipeg Commodity Exchange

Canada's agricultural derivative marketplace (founded in 1887) trades futures and options contracts on barley, wheat, flaxseed, and canola (rapeseed). The WCE demutualized in 2001, a full year ahead of the first U.S. exchange (the Chicago Mercantile Exchange). The WCE converted to full **electronic trading** on December 20, 2004, and thus became the first North American exchange to convert to the new digital format. Canada has always been on the forefront of electronic trading. In the 1960s Canada used the latest technology—the teletype machine—to trade cash hogs. This type of technology needed a new set of trading terms involving trading by description rather than by inspection, which ultimately laid the groundwork for the movement 20 years later away from physical assembly of livestock for cash markets to video and electronic trading. Now the WCE is leading the way for full electronic trading of derivatives. The last open outcry trade occurred on December 17, 2004. Consequently, the WCE is the only North American agricultural derivatives exchange to use a fully functioning electronic trading platform. The Internet address for the WCE is <http://www.wce.ca>.

Others

Derivative exchanges are more numerous now than at any time in history globally. During the last decade, many of the U.S. exchanges have formed relationships with exchanges in other countries to provide off-hour trading opportunities. One new U.S. exchange deserves to be mentioned individually because of the potential for change that it has and likely will have in the future. Eurex US opened for business February 8, 2004, as the first fully electronic derivatives market in the United States Furthermore, it is the first foreign exchange to ask for and get permission to open a new derivatives market in the United States Eurex US is operated by the Deutsche Borse AG and SWX Swiss Exchange, which is the world's largest fully electronic derivatives exchange. Currently Eurex US is trading only financial derivatives. All of the U.S. exchanges have been slow to embrace electronic trading and even in 2005 still relied primarily on physical open outcry auction with lukewarm plans for expansion into electronic trading. Eurex US has got a foothold in the United States now and a proven electronic trading platform that can provide some much needed competition for derivatives trading. The Internet address for Eurex US is <http://www.eurex.com>.

Table 4–3 shows a list of all of the agricultural futures contracts traded on the major North American exchanges. It is interesting to note that the list (current as of June 2005) contains only the major commodities. Yet five years ago the list contained peas, shrimp, and fresh pork bellies just to name a few. Exchanges have had a long history of trying new contracts as long as they feel a need by an industry for the product. A list of all agricultural futures ever traded in the long history of the U.S. and Canadian exchanges would run for several pages and include most agricultural products produced or traded in North America; however, to have staying power futures contracts have to sustain a trading volume to justify trading space at the exchange. To date, only the major commodities can justify such space. This is all the more reason to move to fully electronic trading platforms that don't need bricks and mortar and thus large capital outlays and the variable costs of live traders.

Table 4–3 Major Agricultural and Agriculturally Related Futures Contracts

Contract	Size	Delivery Months
Chicago Board of Trade (CBOT)		
Corn	5,000 bushels	Dec, Mar, May, Jul, Sept
Soybeans	5,000 bushels	Sep, Nov, Jan, Mar, May, Jul, Aug
Soybean oil	60,000 pounds	Oct, Dec, Jan, Mar, May, Jul, Aug, Sep
Soybean meal	100 tons	Oct, Dec, Jan, Mar, May, Jul, Aug, Sep
Oats	5,000 bushels	Jul, Sep, Dec, Mar, May
Wheat	5,000 bushels	Jul, Sep, Dec, Mar, May
Rough rice	2,000 cwt	Sep, Nov, Jan, Mar, May, Jul
SA soybeans (South American)	5,000 bushels	Jan, Mar, May, Jul, Aug, Sep, Nov
Ethanol	29,000 gals	All calendar months
Mini corn	1,000 bushels	Jul, Sep, Dec, Mar, May
Mini soybeans	1,000 bushels	Sep, Nov, Jan, Mar, May, Jul, Aug
Mini wheat	1,000 bushels	Jul, Sep, Dec, Mar, May
Dow jones AIG		
Commodity index	$100 x Index	Jan, Feb, Apr, Jun, Aug, Oct, Dec
Chicago Mercantile Exchange (CME)		
Live cattle	40,000 pounds	Feb, Apr, Jun, Aug, Oct, Dec
Feeder cattle	50,000 pounds	Jan, Mar, Apr, May, Aug, Sep, Oct, Nov
Lean hogs	40,000 pounds	Feb, Apr, Jun, Jul, Aug, Oct, Dec
Frozen pork bellies	40,000 pounds	Feb, Mar, May, Jul, Aug
Class II milk	200,000 pounds	All calendar months
Class IV milk	200,000 pounds	All calendar months
Random length lumber	110,000 board feet	Jan, Mar, May, Jul, Sep, Nov
Butter	40,000 pounds	Mar, May, Jul, Sep, Oct, Dec
Non-fat dry milk	44,000 pounds	All Calendar Months
DAP (fertilizer)	100 tons	Mar, May, Jul, Sep, Dec
VAN (fertilizer)	100 tons	Mar, May, Jul, Sep, Dec
UREA (fertilizer)	100 tons	Mar, May, Jul, Sep, Dec

Table 4–3   (*Continued*)

Contract	Size	Delivery Months
New York Board of Trade (NYBOT)		
Coffee	37,500 pounds	Mar, May, Jul, Sep, Dec
World sugar no. 11	112,000 pounds	Mar, May, Jul, Oct
Domestic sugar no. 14	112,000 pounds	Jan, Mar, May, Jul, Sep, Nov
Cocoa	10 metric tons	Mar, May, Jul, Sep, Dec
Cotton #2	50,000 pounds	Mar, May, Jul, Oct, Dec
Frozen concentrated		
Orange juice	15,000 pounds	Jan, Mar, May, Jul, Sep, Nov
Minneapolis Grain Exchange (MGEX)		
Hard red spring wheat	5,000 bushels	Mar, May, Jul, Sep, Dec
Hard red winter wheat index	5,000 bushels	All calendar months
Soft red winter wheat index	5,000 bushels	All calendar months
Hard red spring wheat index	5,000 bushels	All calendar months
National corn index	5,000 bushels	All calendar months
National soybean index	5,000 bushels	All calendar months
Kansas City Board of Trade (KCBT)		
Hard red winter wheat	5,000 bushels	Jul, Sep, Dec, Mar, May
Winnepeg Commodity Exchange (WCE)		
Canola	20 tonnes	Jan, Mar, May, Jul, Sep, Nov
Feed wheat	20 tonnes	Mar, May, Jul, Oct, Dec
Western barley	20 tonnes	Mar, May, Jul, Oct, Dec
Flaxseed	20 tonnes	Mar, May, Jul, Oct, Dec

Regulating Groups

Commodity Futures Trading Commission (CFTC)

The forerunner to the CFTC was the Commodity Exchange Authority. As futures contracts emerged on financial products and other derivatives emerged such as options and swaps, new **regulations** had to evolve. In 1974 the Commodity Exchange Authority was eliminated and the Commodity Futures Trading Commission was formed. New legislation and changes were added in 1982 with the Futures Trading Act, in 1992 with the reauthorization of the CFTC, and likewise in 2000 with the Commodity Futures Modernization Act. Each of these adjustments allowed for overlapping regulations via the Securities Exchange Commission (SEC) and the Federal Reserve Board (FED) and the ways they have to work together.

The CFTC has the task of regulating all derivative trading including exchanges, traders, and terms of trading for the benefit of the general public. They do not regulate cash and forward contracts except as they relate to manipulation and impacts on derivative markets. The Internet address is www.cftc.gov.

National Futures Association (NFA)

When the CFTC was created in 1974, a provision was added to the act to allow for the creation of futures associations to assist in regulation of the markets. Each exchange develops and enforces its own set of rules and regulations with oversight by the CFTC. The NFA is a self-regulatory group charged with overseeing the market participants that deal with the trading public. As such, all market participants must join and abide by the rules and regulations of the NFA (currently over 4,200 firms and 55,000 individuals are registered). The Internet address for the NFA is www.nfa.org.

Canadian Regulation

Canadian derivatives trading falls under the Commodity Futures Act (CFA) of 1996. Under this law, derivatives trading is regulated by each province and allows for the setting up of Self Regulatory Organizations (SRO). As such the Winnipeg Commodity Exchange has its own SRO which is in turn regulated by the Manitoba Securities Commission. The Internet address for the Manitoba Securities Commission is <http://www.msc.gov.mb.ca>.

CHAPTER 4—QUESTIONS

1. If an arbitrager assumed the normal difference between the March corn futures price and the May futures price was $0.15 per bushel, then saw that the difference had moved to $0.07 per bushel, what action would he take and why?

2. A speculator has observed that the cost of processing soybeans into meal and oil costs on average $0.35 per bushel, yet for the last two years the crush margin has been over $1.00 per bushel. What could the speculator do, if anything, and why?

3. Why are most of the major North American exchanges moving from a mutual organization to a for-profit publicly traded corporation?

4. *Caveat emptor* has long been the mantra of free enterprise markets, yet another term, *caveat venditor*, has emerged. Why?

5. Canadian exchanges are regulated in a much different fashion than U.S. exchanges. What is the major difference?

Fundamentals of Futures Hedging

KEY TERMS		
standardized	bull hedge	under-hedged
liquidity	sell hedge	over-hedged
margin	buy hedge	call contracts
hedging	basis	basis trades
counterbalance	contango	crush hedges
short hedge	backwardation	reverse crush hedge
long hedge	net hedged selling price	selective hedging
bear hedge	net hedged buying price	

Futures contracts have long been the standard for price risk management. This chapter outlines the important concepts and provides examples in the grain, livestock, interest rate, and foreign exchanges markets of how agribusinesses can manage price risks.

OVERVIEW

Most people don't have a problem understanding a simple contract. Two people decide to trade something. One agrees to sell an item and the other agrees to buy it. They agree on when the product will exchange or service will take place, where, and at what price. It really is quite simple. Yet when the term futures contract is mentioned, most people's eyes glaze over and they mutter that they don't understand or that the futures market is just a gambling den in Chicago or New York. But the essence of a futures contract is the same as that for the simple contract. Futures contracts are no more difficult to understand than any other concept in finance or economics.

Once a few simple terms and concepts are understood, anyone can either speculate or hedge with futures contracts. As with all things, people have added (often unnecessarily) complexity and finesse to futures contracts so that a novice immediately thinks that it will take a lot of work to understand how they operate. But the truth is that knowledge of a few basic concepts will open up a powerful risk management tool. The trick is to know the basics really well and ignore all the add-ons—just like with a vehicle. A simple car or truck gets the job of transportation done just as completely as a fully-loaded model of the same make. A fully-loaded model may make a statement about your wealth and may impress others, but getting from point A to point B is done just as well by a basic pickup with a heater and air conditioner as the fully loaded special edition.

Futures Contracts

A futures contract is nothing more than a forward contract that is traded on an organized futures exchange. Special provisions apply, but a futures contract is a strong forward contract that has the bidding for its price open to everyone. A simple forward contract is privately negotiated between two parties, but a futures contract bearing the same terms would

Contract
Chicago Board of Trade (CBOT)
Corn, oats, wheat, soybeans, soybean meal and oil, ethanol, rice, interest rates (Treasury notes, bills, and bonds), equity indexes, gold, and silver
Chicago Mercantile Exchange (CME)
Cattle, hogs, pork bellies, milk, butter, fertilizer, cheese, foreign currencies, equity indexes, and weather
Minneapolis Grain Exchange (MGEX)
Hard red spring wheat and wheat, corn, and soybean indexes
Kansas City Board of Trade (KCBT)
Hard red winter wheat and equity indexes
New York Board of Trade (NYBOT)
Sugar, coffee, cocoa, pulp, orange juice, ethanol, cotton, equity indexes and currencies
New York Mercantile Exchange (NYMEX)
Crude oil, gasoline, propane, heating oil, natural gas, electricity, gold, silver, copper, aluminum, platinum, and petroleum
Winnipeg Commodity Exchange (WCE)
Canola, wheat, barley, and flax

Figure 5–1   Major derivative exchanges and the products they trade

be open to all buyers and sellers. A futures contract is an agreement between the buyer to accept delivery of a product from the seller with prespecified (**standardized**) terms. Buyers and sellers place their bids (the price the buyer hopes to pay) or ask (the price the seller hopes to receive) on the trading floor of the exchange. They must negotiate or find another party with whom to trade. Once a price is agreed upon between buyer and seller, the contract is complete. Since futures contracts are strong, the buyer and seller are free to immediately retrade the contract. The buyer could sell the contract to someone else and be totally out of the obligation. Likewise the seller can buy a contract back from anyone else willing to sell and be out of the promise. This retrading is what attracts risk managers.

Standardization
Figure 5–1 shows the major agricultural commodities that have futures contracts and the specifications for each. Buyers and sellers know exactly what they are getting with a futures contract. Because the contracts don't vary by terms and conditions, the ability to trade out of the contract (retrade) is easy. A buyer can sell or a seller can buy and be in the market and out again without ever going through the delivery process or settlement. The difference between the beginning and ending price determines if a profit or loss was incurred by the trader. Since the contracts are standardized and can be retraded, the futures markets are generally very liquid—traders can get in and out quickly. Futures markets are said to have high **liquidity.**

Leverage

The full value of a contract does not generally have to be advanced to get control of the contract. Usually only a portion of the value, called **margin**, has to be posted to gain control. Margins are set by each exchange and then each brokerage house can add to the minimums if they so choose. The initial margin required to enter a contract is roughly 10 percent of the contract value. For a December corn contract that has a price of $2 per bushel, the contract value would be 5,000 × $2, or $10,000, but the margin to get control of the contract would be approximately $1,000. However, if the contract's value changes, the holder of the contract must pay in any additional losses. Payment for additional losses is called getting a margin call.

Leverage attracts speculators. A cash market speculator would have to pay $10,000 to get control of 5,000 bushels of corn valued at $2 per bushel. However, with futures contracts that same trader could get control of 10 times the volume, or 50,000 bushels, because the amount of margin required is only $1,000 for one contract of 5,000 bushels. Consequently, the futures market speculator could gain 10 times the profit for the same price movements as the cash market speculator (or 10 times the loss).

Those Nasty Margin Calls

To buy a December delivery futures contract for a price of $2 per bushel, the trader must post a margin of approximately $1,000. The margin will have a prespecified level called a *maintenance margin*. Usually the maintenance margin level is approximately 75 percent of the value of the initial margin. A $1,000 margin might have a maintenance margin of $750. The difference between the margin and the maintenance margin level is the amount of money that is allowed to be lost before additional money will be requested—a margin call. The corn contract is for 5,000 bushels and the difference between the margin and the maintenance margin is $250. If the difference between the two margins is divided by the size of the contract, then the trader will know how much can be lost on a per-unit basis. Dividing $250 by 5,000 equals 5 cents per bushel. The trader is allowed to lose up to 5 cents per bushel before a margin call will occur. Table 5–1 shows the effects on margin as price moves.

Notice that as the price moves down the trader is losing money on paper. The trader was allowed to lose up to $250 without a margin call. No margin call occurred until the price had fallen 6 cents per bushel and margin had deteriorated to $700. The margin call is for an amount ($300) to bring the margin account back to full value—$1,000 ($700 + $300).

Table 5–1   Effects of Price Movements on Margin and Margin Calls

Buys one contract (5,000 bushels) of December corn futures at a price of $2 per bushel

Price	Contract Value	Margin Account Balance
$2.00	$10,000	$1,000
$1.98	$9,900	$900
$1.96	$9,800	$800
$1.94	$9,700	$700
Margin call for $300		$1,000
$1.92	$9,600	$900

Table 5–2   Effects of Price Movements on Margin and Paper Profits

Price	Contract Value	Margin Account Balance
$2.00	$10,000	$1,000
$2.02	$10,100	$1,100
$2.04	$10,200	$1,200
Profit withdrawal $200		$1,000
$2.03	$10,150	$990

Traders hate to get margin calls because they signify that trades are losing money. The idea behind a margin call is that because the trade was allowed with only a fraction of the contract value posted, losses need to be paid in as they occur to assure financial performance from the trader. The purpose of the maintenance margin level is to allow a certain amount of price movement that will naturally occur in markets so that a margin call is not made for every price movement.

If a trade is generating paper profits then the trader can withdraw the extra revenue. The converse of a maintenance margin for profits is not in widespread use, but each brokerage house will set an amount of paper profit that has to be made before they will allow for it to be withdrawn. Table 5–2 indicates how a paper profit is handled.

At this point the trader could withdraw $200 in paper profits. If the trader in the example in Table 5–2 withdraws the $200 in paper profits, the margin account balance falls back to the $1,000 initial requirement level.

Margin is important to understand because it is a cash flow financial consideration for businesses. To hold a futures position either as a speculator or hedger is to be subject to margin calls. For paper losses, businesses will have to inject cash to hold the position and for paper gains, cash can and should be withdrawn and managed from a time value of money standpoint.

Concept of Counterbalance

Hedging is the process of counterbalance—one action is offset by another action. A single action is simply speculation, complete with all the risks associated with the action. Hedging is an attempt to cover some aspect of the risk of the speculative action with another action. For hedging to work, the results of the two or more actions must be opposite, otherwise risk has actually increased rather than decreased.

Futures contract hedging entails having opposite cash and futures positions. If a grain elevator buys corn and stores grain for a certain period of time they have the risk that prices will fall during the holding period. What they need is a financial product that gains in value when cash corn prices fall to offset the loss of value in the cash market. When they buy corn initially, they could simultaneously sell a corn futures contract. If corn prices fall, their cash corn loses value, but the sell position in the futures gains in value because it can be repurchased at a lower price, thus, the hedge. What is lost in one market is gained in another. The hope is that the losses exactly cover the gains.

What if the elevator had simultaneously bought a futures contract when they purchased the cash corn? Now they would have two simultaneous positions in different markets, but they are not counterbalanced. The two markets will move together, not oppositely.

Table 5–3   Effects of Counterbalance and Parallel Positions

Cash Position	Counterbalance Futures Position	Parallel Futures Position
Buys corn @ $2.20	Sells corn futures @ $2.30	Buys corn futures @ $2.30
Price Decrease		
Sells corn @ $2.00	Buys corn futures @ $2.10	Sells corn futures @ $2.10
Loss of $0.20	*Gain of $0.20*	*Loss of $0.20*
Price Increase		
Sells corn @ $2.40	Buys corn futures @ $2.50	Sells corn futures @ $2.50
Gain of $0.20	*Loss of $0.20*	*Gain of $0.20*

Hedging must involve **counterbalance.** If it does not, then it is not a price risk management tool, but rather another form of speculation. Table 5–3 reveals the effects of counterbalance in a hedge and a parallel speculative position with both a price increase and decrease.

The counterbalanced position always has the opposite results of the cash position while the parallel position always has the same effect. Counterbalancing is hedging while parallel positions are always a doubling of the risk of the cash position. Parallel positions will yield twice the gain and also twice the loss as the example in Table 5–3 shows; they double your pleasure when they work in your favor but double your sadness when they don't.

This trade-off is critical to understanding and using hedging. If the hedge is not properly placed, then price risk is not managed, but in fact may, as parallel positions show, double the exposure to price risk. Speculation is not a bad or wrong thing to do; it becomes a problem when a business wants and needs to hedge, but in fact is entering into a higher level of risk, not a lower level.

Only Two Types of Hedges

There really are only two ways to hedge. An initial sell position in the futures is called a **short hedge,** as shown in the example in Table 5–3. A short hedge is used to protect against declining spot market values. The other type of hedge is an initial buy position in the futures market called a **long hedge.** Long hedges are used to protect against increasing values in the cash market. That's it. Almost all hedges for price risk management will be a simple short or long hedge. More complex futures hedging exists, but the structure is just a combination of various types of short and long hedges.

The example in Table 5–3 of a perfect short hedge reveals the purpose: to protect against the spot price going down, which it did. When the cash corn price decreased, the short futures position gained in value. Short hedges are widely used in agriculture by producers who have growing crops or livestock and agribusinesses such as elevators (as in Table 5–3) that need to protect an asset (owned corn) against a declining value.

Table 5–4   Long Ingredient Hedge for a Food Manufacturer

Cash Market	Futures Market
June 1, sells finished product Wheat price set @ $3.40/bushel	Buys July wheat @ $3.50/bushel
June 8, buys wheat @ $3.60/bushel	Sells July wheat @ $3.70/bushel
Loss of $0.20/bushel	Gain of $0.20/bushel

Some agricultural businesses need protection against rising prices. A food company that uses wheat to manufacture a food product will price the finished food product (bread or pancake mix, for example) to wholesalers in advance but will then have to go into the spot market regularly, maybe even each week, to buy wheat for the factory. Once it fixes its final price to the wholesaler, the food company has the risk that the raw ingredients it purchases will increase in price and thus squeeze profit margins. Table 5–4 shows how they could long hedge that risk.

The food company suffered a loss of 20 cents per bushel on the cash wheat, but the hedge offset the loss with an exact gain. The company, in effect, bought the wheat at the original $3.40 per bushel.

A hedge is called long or short based on the initial position in the futures market. Short hedges are also called **bear hedges** or **sell hedges** and long hedges are called **bull hedges** or **buy hedges.** The most generally used terms in the industry, however, are short and long. Short and long are also used in reference to the cash position. If a business has forward sold a product in the spot market, they are said to be *short the cash*. Conversely, if they have purchased a product in the spot market they are said to be *long the cash*. Short and long, for both futures and cash positions, refer to initial positions only. If a business is long the cash (i.e., an initial buy in the spot market) when they sell the product, the action of getting out of the long cash position is not called being short, it is referred to as a sell.

The Role of Futures Contracts in Hedging

Futures contracts derive their value from the underlying cash market. The same factors that cause the cash market price to change will result in a change in the futures contract price. The two prices will move in tandem, or as the trade says, the two prices "tend to trend together." This derivative aspect of the two prices is the reason that futures contracts can be used as a counterbalance for cash positions.

Absolute Price Movements

When the price of an item changes in the cash market, that price activity is called an absolute price movement. Figure 5–2 demonstrates the movement for the largest grain market product—corn. Figure 5–2 shows the prices of corn in one of the United States' largest corn producing states, Iowa. The motion of the prices is quite pronounced in terms of highs and lows for the year, i.e., the absolute price movement. The high for the year was $1.90 per bushel and the low was $1.70. The $.20 difference represents a price movement

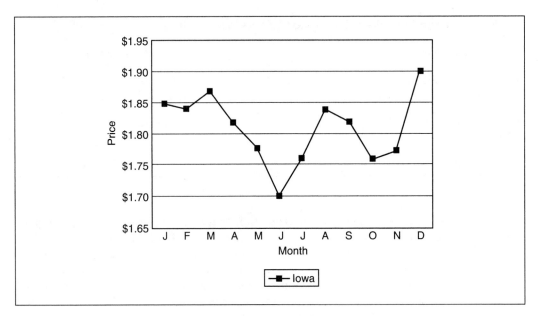

Figure 5–2   Iowa cash corn prices, 2001

from the highest to the lowest point of the year. These absolute price movements are the source of the price risk that producers and agribusinesses face in dealing with the products. If a business is not going to hedge with futures contracts, then knowledge of the absolute price movement for the cash commodity needs to be studied so that other risk management tools can be used, such as timing of sales or processing or insurance.

Historical price information can be used to calculate certain statistical values such as means, medians, and variances. These values can be helpful in analyzing how risky the movements are to the business. Certain commodities exhibit patterns that are seasonal (within a calendar or growing period) or cyclical (over several calendar years or growing periods). It is important to know the likely tendencies of price movements as a way to help manage the risk of the price movement. Later in this book specific characteristics of price patterns will be incorporated in risk management hedging strategies for grains and livestock.

Just as cash prices have their own absolute price movements, futures contract prices have their own absolute price movements. Even though the futures price is derived from the cash price, futures contracts are standardized and relate to future prices rather than current-day prices and thus don't exactly match the spot mandate for the cash commodity. Because futures contracts are traded on exchanges and price is determined by open outcry auction, prices exist for each day the commodity is traded and for each moment within the day. In fact, futures prices are more readily available than most cash market prices merely because they are centralized. Figure 5–3 illustrates the futures price movement for corn for the year 2001. Futures prices exhibit price patterns and can be analyzed historically for statistical properties such as means and variances.

Relative Price Movements

Hedging necessitates counterbalance, thus both cash and futures absolute price movements will exist side by side. The importance of price movements with a hedge is not the absolute price movement of either the cash or futures but how the two prices relate to each other

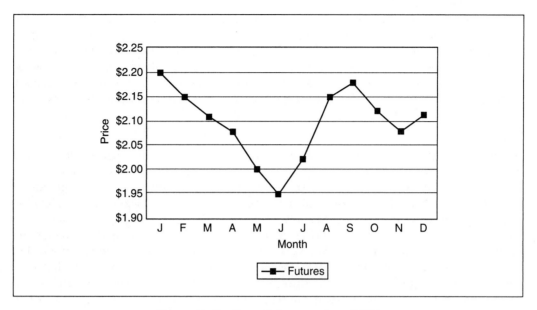

Figure 5–3   Corn futures prices, 2001

during the hedging period, that is, the relative movement. The two price series will move in tandem, but because each price movement has unique characteristics, they will not move exactly the same. When a hedge is placed, only the relative differences between the two absolute price movements has importance. The two price movements counterbalance, but not perfectly. *Hedging removes the risk of absolute price movements and replaces it with the risk of relative price movement.* Figure 5–4 shows the relative price movement of cash

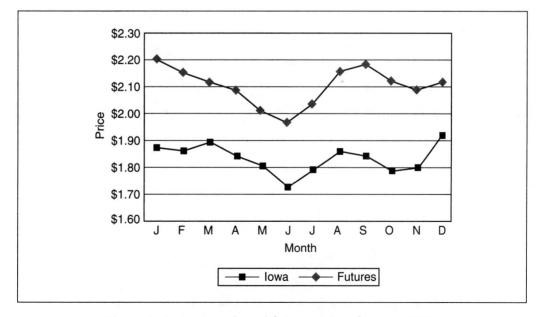

Figure 5–4   Iowa cash and futures prices for corn, 2001

corn prices and corn futures prices. The two markets tend to trend together, but not perfectly. Hedging can manage the risk of an absolute cash price movement only by substituting that risk with the risk that the cash and futures markets will not move perfectly together. The difference between the cash market price and the futures market price is called **basis.** Basis is the risk of relative price movements. The general statement is now modified to read: *Hedging removes the risk of absolute price movements and replaces it with basis risk.*

Basis

A basis value exists for every cash market. Futures prices are centralized but cash prices exist wherever individuals or businesses want to trade. Corn in Iowa is not the same as corn in Nebraska, even though they border each other, nor is it the same as in Texas. Not only do they differ by location, but the corn will be used differently at each location and will exhibit different price levels and patterns as shown in Figure 5–5. Nebraska and Iowa prices were almost identical from January to May 2000, but started to change during the summer and fall. The Texas price followed the same general price pattern as in Iowa and Nebraska but differed radically most of the year. When the futures price for corn is added, the basis for each market emerges and would be different because the cash prices are different, as Figure 5–5 clearly shows. Basis values for one cash market cannot generally be substituted for basis values for another. The importance of basis values will be more easily understood when hedging is more fully explained. Basis patterns are important and are discussed in more detail later in this chapter. Suffice to say at this point, basis is the most important aspect of hedging to understand.

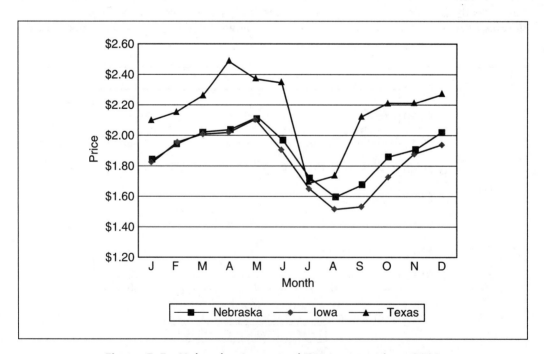

Figure 5–5   Nebraska, Iowa, and Texas corn prices, 2000

Basis is defined as the difference between the futures price and the cash price, or

basis = futures price (*Fp*) at time *t* less the cash price (*Cp*) at time *t*,

or

$$B = Fp_t - Cp_t$$

Markets that have futures prices that are higher than the cash price are said to be in **contango,** while markets that have the cash or spot price higher than futures prices are in **backwardation.** Using the above formula, markets in contango will produce a basis that is positive and markets that are in backwardation will have a negative basis.

General Rules and Effects of Hedging with Futures

Perfect Hedges

Perfect hedges are like perfect humans—they don't exist nor will they ever exist, but they can serve as a benchmark. *A perfect hedge eliminates the cash market risk and has no basis risk effects.*

A grain elevator buys 5,000 bushels of cash corn for a price of $2.00 per bushel. The elevator simultaneously sells a March delivery corn futures contract for $2.20 per bushel. Two days later the elevator sells the cash corn for a price of $1.95 per bushel and lifts the hedge by buying a corn futures contract for $2.15 per bushel. Table 5–5 reveals the effects of the hedge. The cash price risk to the elevator was that the price of corn would decrease and the corn would have to be sold for a lower price than the purchase price. The sell position in the futures provided the counterbalance. When the cash price decreased the futures price did likewise by the same amount, thus a perfect trade-off. The elevator did not make any money on the deal, but didn't lose any either. Had they not hedged, the loss would have been 5 cents per bushel in the cash market.

If prices had increased instead, then a gain in the cash market would have been offset by a loss on the futures position as shown in Table 5–6. No matter whether prices go up or down, the elevator ultimately gets the same selling price for the corn because the gain or loss in either market exactly matches the other. The elevator sells the corn for $2.00 per bushel net regardless of whether or not prices went up or down. The elevator paid $2.00 per bushel for the corn, so the deal yielded no net gain or loss.

Table 5–5   Perfect Corn Hedge, Cash Price Decrease

Cash Market	Futures Market	Basis
February 1, buys corn @ $2.00/bushel	Sells one March corn futures @ $2.20/bushel	$0.20
February 3, sells corn @ $1.95/bushel	Buys one March corn futures @ $2.15/bushel	$0.20
Loss of $0.05/bushel	Gain of $0.05/bushel	

Table 5–6   Perfect Corn Hedge, Cash Price Increase

Cash Market	Futures Market	Basis
February 1, buys corn @ $2.00/bushel	Sells one March corn futures @ $2.20/bushel	$0.20
February 3, sells corn @ $2.05/bushel	Buys one March corn futures @ $2.25/bushel	$0.20
Gain of $0.05/bushel	Loss of $0.05/bushel	

Without hedging, the elevator would have lost 5 cents per bushel if prices moved down, but would have made 5 cents per bushel if prices had moved up. The risk for the elevator is not that prices will increase, but rather that they will decrease. If the elevator gets in the market and places the hedge and if beginning basis remains the same as when the elevator gets out of the market and lifts the hedge ending basis the hedge will be perfect—neither a gain nor a loss will occur. Beginning basis (basis when the first trade occurred) will equal ending basis (basis when the offsetting trade occurred).

If the elevator has a perfect hedge then the effect is no gain or loss on the transaction. The elevator would have been just as well off by not entering the deal at all. In fact, when brokerage fees and the time value of money for the margin are considered, the elevator actually lost money by hedging with a perfect hedge versus not doing the cash deal at all. Therein lies the paradox—a perfect hedge is not desirable. *What hedgers really want and need are imperfect hedges.*

Imperfect Hedges

Imperfect hedges have a beginning basis and an ending basis that are different. A change in the basis will result either in a net loss or gain for the hedger. Table 5–7 shows a basis change that results in a net gain for the hedger, regardless of whether prices increase or decrease. *Price direction is unimportant, only the relative movements between the cash and futures markets—basis—matters.*

When a net gain from basis movements occurs the basis is said to have improved. A deterioration in the basis will result in a net loss for the hedger. Table 5–8 is an example of a basis that deteriorates. The elevator lost 2 cents per bushel regardless of whether price increases or decreases.

Imperfect hedges have gains and losses in the cash and futures markets when the prices go up or down. *The only thing that matters is the net effect of both markets together.* It is important to notice that the net gain or loss exactly matches the change in the basis. In both examples in Tables 5–7 and 5–8 the net gain and loss is 2 cents per bushel which is exactly the change in the basis between the beginning basis and the ending basis.

For long hedges the results are just the opposite. Table 5–9 shows an example of a feed company that forward sells mixed feed for delivery in two weeks with corn prices set at $2 per bushel. The feed company does not have the corn in storage and must buy the corn in the spot market a day before delivery and manufacture the feed. The feed company has the risk that corn prices will increase and squeeze or eliminate its profit margin in the feed.

The example in Table 5–9 shows the feed manufacturer losing 20 cents in the cash market, but the futures hedge made 22 cents because the basis improved 2 cents. The feed

Table 5–7 Basis Improvement, Short Hedge

Cash Market	Futures Market	Basis
February 1, buys corn @ $2.00/bushel	Sells one March corn futures @ $2.20/bushel	$0.20
Price Decrease		
February 3, sells corn @ $1.95/bushel	Buys one March corn futures @ $2.13/bushel	$0.18
Loss of $0.05	Gain of $0.07	$0.02 Change
Net gain of $0.02($0.07 − $0.05)		
Price Increase		
February 3, sells corn @ $2.05/bushel	Buys one March corn futures @ $2.23/bushel	$0.18
Gain of $0.05	Loss of $0.03	$0.02 Change
Net gain of $0.02($0.05 − $0.03)		

Table 5–8 Basis Deterioration, Short Hedge

Cash Market	Futures Market	Basis
February 1, buys corn @ $2.00/bushel	Sells one March corn futures @ $2.20/bushel	$0.20
Price Decrease		
February 3, sells corn @ $1.95/bushel	Buys one March corn futures @ $2.17/bushel	$0.22
Loss of $0.05	Gain of $0.03	$0.02 Change
Net loss of $0.02($0.03 − $0.05)		
Price Increase		
February 3, sells corn @ $2.05/bushel	Buys one March corn futures @ $2.27/bushel	$0.22
Gain of $0.05	Loss of $0.07	$0.02 Change
Net loss of $0.02($0.05 − $0.07)		

Table 5–9 Basis Improvement, Long Hedge

Cash Market	Futures Market	Basis
February 1, forward sells Feed corn price set @ $2.00/bushel	Buys one March corn futures @ $2.20/bushel	$0.20
February 14, buys corn @ $2.20/bushel	Sells one March corn futures @ $2.42/bushel	$0.22
Loss of $0.20/bushel	Gain of $0.22/bushel	$0.02 Change

Net gain of $0.02 ($0.22 – $0.20)

manufacturer made its profit margin plus an extra 2 cents per bushel on the corn because they had an improving basis on its hedge.

General rules of thumb about basis movements can be established. Short hedgers want the basis to narrow, that is, they want the ending basis to be less than the beginning basis. However, long hedgers want the basis to widen, as the example in Table 5–9 shows.

Net Hedged Prices

In the previous examples of Tables 5–7 and 5–8, a grain elevator buys spot corn and then hedges. A grain elevator is buying spot corn and later selling, hopefully at a profit. When the elevator buys the spot corn, the buying price is set ($2.00 per bushel in the previous examples). The elevator later sells the corn in the spot market, but the cash selling price is not the final net selling price because the effects of the futures market transactions have to be calculated as well. The net effect of both cash and futures market transactions yields the net selling price for the hedge.

Table 5–8 shows a net loss when price decreases 2 cents per bushel. The elevator paid $2.00/bushel for the corn and had a net loss of 2 cents per bushel on the sale, so the actual net selling price is $1.98 per bushel considering the effects of both cash and futures transactions. The elevator sold the corn in the cash market for $1.95 per bushel, but made 3 cents per bushel from the futures transactions. In effect then, the elevator had a **net hedged selling price** of $1.98 per bushel.

The formula is

NHSP = FCSP + NF

where

NHSP = net hedged selling price

FCSP = final cash selling price

NF   = net futures gain/loss

Using the formula for the results in Table 5–8,

Price decrease

NHSP = $1.95 + $0.03 = $1.98

Price increase

NHSP = $2.05 + (–$0.07) = $1.98

The elevator paid $2 per bushel for the spot corn and ended up selling it for a net price of $1.98 per bushel which resulted in a 2-cents per bushel loss (exactly the change in the basis).

The long hedge example in Table 5–9 illustrates that the feed manufacturer has forward sold feed for delivery in two weeks with the price of corn fixed at $2 per bushel. When it actually buys the corn in 2 weeks it has to pay $2.20 per bushel in the spot market. Yet the futures transactions made 22 cents, thus the feed company had a gain of 2 cents. When the feed company forward sold the feed for delivery in two weeks it assumed a buying price of corn at $2 per bushel. It paid $2.20 in the spot market, but 22 cents in the futures, thus it really bought the corn for a **net hedged buying price** of $1.98 per bushel.

The formula is

NHBP = FCBP – NF

where

NHBP = net hedged buying price

FCBP = final cash buying price

NF   = net futures gain/loss

Using the formula for the results in Table 5–9,

NHB   = $2.20 – $0.22 = $1.98 per bushel

Using Futures in the Grains and Oilseed Markets

The oldest futures-type contracts in the United States originated in the grain markets. Needless to say, much has been written and observed about grain futures contracts. The range of traders in grains is perhaps the widest of all commodities. The list begins with simple short hedges for wheat farmers and continues up in complexity to large trading giants like Cargill that use complex **basis trades** to move grain around the world.

Wheat, corn, oats, barley, soybeans, canola (rapeseed), and flaxseed are crops that provide food for humans and animals. More and more, though, industrial uses are being made of these crops, such as ethanol, biodiesel, and plastics. Pharmaceuticals are becoming more mainstream in crop agriculture as ethnobotanists and biotech scientists discover products within plants that have medicinal properties. Higher value uses for major crops will necessitate a better understanding of how to manage the risk of price changes.

Basis Factors

Grains and oilseeds are produced and harvested on a fairly regular schedule. Different varieties mature earlier and later and widen the harvest season, but with the exception of winter wheat and barley crops, all major grain and oilseed crops are planted in the spring and harvested in the fall. Grains and oilseeds are storable for extended periods of time. The combination of the biological process of planting and harvesting on a fairly fixed time schedule and the storable characteristics of the product provide a theoretical underpinning to develop a pattern for seasonal price movement. The price at harvest should be the lowest during the year as all of the year's product is available to the market. As the product is placed in storage and pulled out during the year for use, the cost of storage will be reflected in the price—that is, after harvest the price should gradually increase to reflect the cost of carry. At the beginning of this chapter, Figure 5–2 illustrated this principle with the seasonal average cash price for corn in Iowa. The price is lowest at harvest in the early fall and gradually increases throughout the spring and then begins to fall as harvest approaches again. The model is not perfect, of course, but it does provide insight into the seasonal movement for grain and oilseed prices.

Cost of Carry Model

Storable products should have the cost of carry reflected in the price at various points in time. The cost of carry is the term used to reflect not only the physical cost of storage but the financial costs as well. Financial costs include the cost of insurance and should also contain the opportunity cost of money. Storage also has a deterioration cost because there is a risk that the product will change during storage and lose value or be totally lost. The cost of carry model is

$$P_{t+1} = P_t + COC_t$$

where

P_t = price in time period t

P_{t+1} = price in next time period

COC_t = cost of carry from one time period, t, to the next time period, $t+1$

For a grain like corn, this model works fairly well, as Figure 5–6 illustrates, from mid fall to early summer. As the new crop is harvested, P_t reflects the new supply and demand conditions

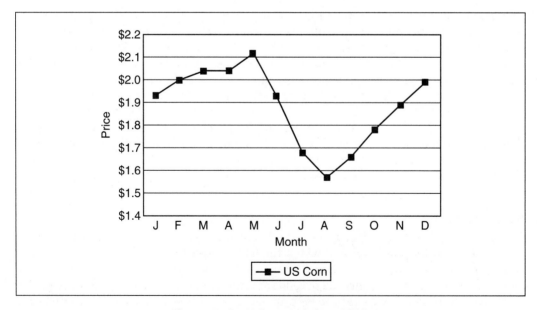

Figure 5–6   U.S. corn price, 2000

and restarts the seasonal pricing. If the model is started in September (P_t), then the price in October (P_{t+1}) should be the price in September plus the cost of carry for one month. The price in November should be the October price plus the cost of carry for another month and so on and so forth until the new crop starts being harvested in summer and then there is a new P_t as reflected by supply and demand for that particular crop year.

Basis and Cost of Carry

Basis is the numerical difference between a futures price and a cash price. If the cash price in Chicago for corn on November 1 is $2.00 per bushel, what price should a futures contract for December delivery trade for relative to the November 1 price? A rational value would be the cost of carrying the corn from the time period November 1 to December 1. The expected value of the futures contract would be the cash price in Chicago plus the amount of one month's cost of carry. As December approaches the cost of carry lessens, and in December the cash price and the December futures price should be approximately the same. Figure 5–7 shows the cash price and futures price moving together as the futures contract nears maturity, reflecting the conceptual nature of the cost of carry model. The Iowa cash price for corn and the July delivery futures contract price differed by 60 cents per bushel in October 2001, but by May 2002, the difference had fallen to only 20 cents per bushel. As time eroded, so did the value of the cost of storage. The cash and futures prices for storable commodities should gradually come together (a lessening of the basis) as the futures contract nears maturity.

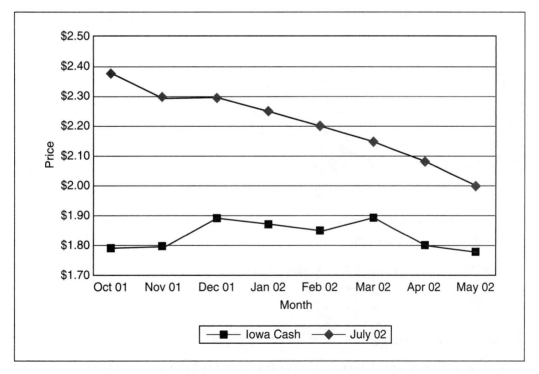

Figure 5–7   Iowa cash corn price and July 2002 corn futures price

Basis and the Cost of Transportation

In Figure 5–7 the cash price of corn in Iowa and the futures price are moving closer together as the contract matures. But Iowa corn and corn futures are not the same because the futures contract is standardized for delivery to Chicago, St. Louis, or Toledo at a future date. As the Iowa cash price and futures price difference gets less due to the eroding nature of cost of carry, the prices should not be the same in July when the futures contract matures. The price of Iowa cash corn in July and the July futures contract for delivery elsewhere should differ by the cost of transportation from Iowa to one of the three delivery points. Transportation costs are fairly stable and thus the cost of transportation can be treated like a constant value throughout the holding period for the futures contract. The grain trade treats the transportation cost as a constant value to either be added or subtracted, depending on whether they are buying or selling the product, to the cash price relative to the futures price. For example, Tulsa, Oklahoma, can be reached by grain barges via rivers and canals from the Mississippi River. This added cost of transportation is called the "Tulsa basis," implying the cost of transportation by river grain barge from the Mississippi River to the grain port in Tulsa.

Grain and Oilseed Basis Model

The basis for grain crops and oilseed crops can be expressed now as

$$B_t^{t+n} = F_{P_t}^{t+n} - C_{P_t}$$

where

$\quad B_t^{t+n}$ = basis at time t for delivery in time $t + n$

$\quad F_{P_t}^{t+n}$ = futures price at time t for contract month $t + n$

$\quad C_{P_t}$ = cash price at time t

$\quad (t + n)$ = futures delivery month

The estimated basis is

$$\hat{B}_t^{t+n} = C_{P_t} + \sum_{t}^{t+n} COC_t + T$$

where

$\quad \hat{B}_t^{t+n}$ = estimated basis at time t for delivery at time $t + n$

$\quad C_{P_t}$ = cash price at time t

$\quad COC_t$ = monthly cost of carry

$\quad T$ = transportation cost

The estimated basis model can be useful to determine how the real basis is conforming to the basis model of cost of carry and transportation. Estimates of basis values are very important to hedgers.

Production Hedge

Farmers are involved in a complex process of combining inputs, managing those inputs in a crop-growing process, and harvesting the results. Once they commit to a growing a crop, farmers do not know with certainty what price they will get several months later at harvest.

Table 5–10 Wheat Farmer Hedging Example with a Price Decline

Cash	Futures	Basis
Aug 1, plants 1,000 acres Estimated production of 30 bushels/acre (30,000 total) Local cash price $4.00/bushel	Sells 6 July Kansas City wheat futures @ $4.40/bushel (5,000 bushels per contract with 6 contracts =30,000 total)	$0.40
June 1, harvests 32,000 bushels and sells locally @ $3.20/bushel for a total revenue of $102,400	Buys 6 contracts @ $3.40/bushel	$0.20
	Net gain of $1.00/bushel ($30,000 total)	Change of $0.20

Net hedged selling price = $4.1375/bushel
($102,400 + $30,000 = $132,400/32,000 = $4.1375)

As the production process proceeds and the price of the product decreases (remember that price tends to decline as harvest approaches), a hedge would need to increase in value to offer the proper counterbalance. Thus production hedges need a futures position that is short initially so that if price decreases, a gain will occur with the futures position. *Regardless of the type of product or time period involved, a production price risk will always be hedged short.*

Wheat Producer Example

A Kansas wheat farmer plants winter wheat in late summer or early fall and harvests the next calendar year in late spring or early summer. The farmer normally plants 1,000 acres and her average yield over the last five years has been 30 bushels per acre. She estimates this year's crop at 30,000 bushels. She plants the crop in late August. The local cash price for hard red winter wheat is $4.00 per bushel. She harvests her crop in June and has a total crop of 32,000 bushels. She received a price from her local cash market of $3.20 per bushel. When she planted the crop she also hedged by selling six Kansas City July futures contracts at a price of $4.40 per bushel. When she sold her cash crop in June she bought back her futures contracts at a price of $3.40 per bushel. Table 5–10 summarizes the results and shows the net hedged selling price. Notice the net hedged price had to be adjusted because the cash and futures quantities were not a perfect match. A quantity adjustment factor must be applied to the futures portion to use the formula for net hedged prices. The quantity adjustment would be the futures quantity divided by the cash quantity. In the example, that would be 30,000/32,000 = 0.9375. This quantity multiplier would then be applied to the amount per bushel that the futures positions made ($1.00), or $0.9375 per bushel. The futures position made 93.75 cents per bushel for the cash position of 32,000 bushels. Using the formula for a net hedge selling price (NHSP),

$$NHSP = \$3.20 + \$0.9375 = \$4.1375$$

This example has the farmer in an **under-hedged** situation by 2,000 bushels. The farmer estimated the crop to be 30,000 bushels and hedged accordingly. However, the actual crop

Table 5–11   Wheat Farmer Hedging Example with a Price Increase

Cash	Futures	Basis
Aug 1, plants 1,000 acres Estimated production of 30 bushels/acre (30,000 total) Local cash price $4.00/bushel	Sells 6 July Kansas City wheat futures @ $4.40/bushel (5,000 bushels per contract with 6 contracts =30,000 total)	$0.40
June 1, harvests 32,000 bushels and sells locally @ $4.50/bushel for a total revenue of $144,000	Buys 6 contracts @ $4.70/bushel	$0.20
	Net Loss of $0.30/bushel ($9,000 total)	Change of $0.20

Net hedged selling price = $4.21875/bushel
($144,000 + $9,000 = $135,000/32,000 = $4.21875)

turned out to be 32,000 bushels. The futures contracts are standardized and thus inflexible; therefore, this situation is very real. The futures and cash quantity positions will almost never be equal and a quantity adjustment multiplier must be used.

This example also shows a narrowing of the basis from the beginning of 40 cents per bushel to 20 cents at the end for a net gain of 20 cents. The local cash price when the crop was planted was $4.00 per bushel. The farmer did not have the product to sell at that time so she could not take advantage of that price. When the farmer sold her grain several months later, the price had fallen to $3.20 per bushel, a price decline in the spot market of $0.80 per bushel. However, the futures position gained $1.00 per bushel during the same time because the futures price declined by $1.00 per bushel, thus the 20-cent improvement in the basis ($1.00 − $0.80). The farmer didn't get the net effect of the 20 cents because the cash and futures quantities were mismatched. The farmer's net futures gain adjusted for the quantity difference was $0.9375 per bushel. So the farmer wanted protection against the 80 cents the cash price moved, but instead got the gain in the basis such that she received $0.9375. Therefore, the farmer was shooting for the $4.00 per bushel original cash price, but got instead $4.1375. The $0.1375 difference is also (and always is) the difference between the cash price movement ($0.80) and the quantity adjusted futures price movement ($0.9375).

A production hedge is the same regardless of whether the product is cotton, canola, or soybeans. The core concept of a production hedge is that the risk of a declining price during the production process needs to be counterbalanced with a financial asset that gains in value. The financial asset in the example in Table 5–10 is a short futures position. The wheat farmer in the example not only received price risk protection, but actually got a bonus of almost $0.14 per bushel because of a favorable basis change.

What about a price increase? The wheat farmer would have had a gain in the cash price but would have a loss in the futures side. Table 5–11 exhibits that effect.

Using the net hedged price formula,

$$\text{NHSP} = \$4.50 + (\$0.30 \times 0.9375)$$

$$\text{NHSP} = \$4.50 + (\$0.28125) = \$4.21875$$

The Decision to Over- or Under-Hedge

The farmer was under hedged by 2,000 bushels. When cash prices moved lower (against the farmer) under hedging lowered the net hedge price (the futures gain was $1 per futures bushel, but was lowered to $0.9375 per actual cash bushel). However, when prices moved up (in favor of the farmer) under hedging increased the net hedge price (the futures loss was 30 cents per futures bushel, but was lowered to $0.28125 per actual cash bushel).

This production hedge can serve to develop the rules of thumb for over and under hedging. *A hedger should under hedge if he believes there is a stronger probability of cash prices moving in his favor than against him. He should over hedge if the chance that cash prices will move against him is greater than in his favor.*

Storage Hedge

Grain and oilseed crops can be stored for long periods of time if proper conditions are maintained. A large segment of the grain and oilseed industry is composed of merchants that perform the task of storing grains and/or oilseeds. Often these market participants are also performing other functions such as assembly, grading, and transporting. The classic example is a merchant called a grain elevator. Elevators buy and sell grains and oilseeds from producers. They will commingle the grains and oilseeds to form larger lots or for a change in grades. Elevators range from simple operations in farming areas that just gather the crops and put them in larger lots for train or truck shipment to terminal centers on major rivers, rail lines, or road junctions that buy and sell from farmers and from other country elevators. Regardless of the size or location of the elevator, they all have a storage function to perform. Grain and oilseed crops are purchased and stored for later sale.

Elevators that hold crops for later sale run the risk that prices will decline after they purchase the grain or oilseed. Storage hedges are short hedges.

Grain Merchant Storage Hedge

A grain merchant buys 5,000 bushels of corn on March 15 for a price of $2.10 per bushel. They hedge it by selling one May corn futures contracts at a price of $2.25 per bushel. On April 15 the merchant sells the cash corn for $2.00 per bushel and lifts the hedge at $2.10 per bushel. Table 5–12 summarizes the results.

Table 5–12 Grain Merchant Storage Hedge

Cash	Futures	Basis
March 15, buys corn (5,000 bushels) at $2.10 per bushel	Sells one May corn at $2.25 per bushel	$0.15
Grain in storage		
April 15, sells corn at $2.00 per bushel	Buys one May corn at $2.10 per bushel	$0.10
Loss of $0.10 per bushel	Gain of $0.15 per bushel	Change of $0.05
Net hedged selling price = $2.00 + $0.15 = $2.15 per bushel		

The merchant purchased the grain for $2.10 per bushel and ended up selling it for a net price of $2.15 per bushel for a net difference of 5 cents per bushel. What does the net difference of 5 cents per bushel represent? Using the cost of carry concept, the difference should represent approximately the cost of storage for the month that the merchant held the corn. To the extent that the actual costs of storage were less than 5 cents, the merchant had a net gain. If the storage costs were more than 5 cents the merchant had a net loss.

Since grains and oilseeds are storable, the estimated change in the basis values should narrow (decrease) over time. This allows for an opportunity for merchants when they hedge the grain and oilseed to be able to cover the cost of storage if the basis does in fact narrow.

Forward Pricing Hedging

Grain merchants, food processors, and feed processors have the opportunity to forward price grain and oilseeds. A food processor that is using wheat to make a breakfast cereal product generally has to forward sell the final food product to wholesalers before it is manufactured. To properly arrive at a price for the final product, an assumption about what the ingredient prices will be has to be made. The food manufacturer will have to forward price the final product and, consequently, in effect forward price the ingredients such as wheat at the same time. The food processor's risk is that the ingredient price will increase before they can purchase the item and get the final product produced.

Food Manufacturer Hedge

A cereal manufacturer forward prices the product to a large wholesaler at $3.00 per item. The breakfast cereal product will be delivered to the wholesaler in two weeks. The food manufacturer will buy the ingredients two days prior to delivery, manufacture the product, and then deliver so that the product is fresh. When the food manufacturer sets the $3.00 per item price, they assumed the wheat could be purchased for $3.50 per bushel. The food processor has, in effect, forward sold the wheat at a price of $3.50 per bushel. The risk is that from the pricing time until the actual wheat is produced, the price of wheat will increase and eliminate the processor's profit margin or severely squeeze the margin. The processor needs a counterbalance of a financial asset that would gain in value when wheat prices in the cash market increase. The processor needs to be a long hedger. Table 5–13 illustrates the hedge.

The processor sells the wheat for $3.50 per bushel and ends up buying it for a net price of $3.53 per bushel. The manufacturer had his profit margin reduced by 3 cents per bushel.

Table 5–13   Food Manufacturer Hedge

Cash	Futures	Basis
Nov 1, forward sells wheat at $3.50 per bushel	Buys one December wheat contract at $3.65 per bushel	$0.15
Nov 12, buys wheat at $3.60 per bushel	Sells one December wheat contract at $3.72 per bushel	$0.12
Loss of $0.10 per bushel	Gain of $0.07 per bushel	Change of $0.03
Net hedged buying price = $3.60 − $0.07 = $3.53 per bushel		

The hedge protected the processor, but not perfectly because the basis moved against the manufacturer by 3 cents per bushel. Yet if the processor had not hedged at all, the loss would have been the full 10-cent difference between the forward selling price and the buying price.

Basis Traders and Basis Contracts

The examples for production, storage, and forward pricing hedges show the effects that basis movements have on the net hedged prices. Once a hedge is properly placed, all that really matters is the movement in the basis—that is, the relative price movements between the cash price and the futures price. As hedgers become more familiar with the process and outcomes from hedging, basis develops into a more important factor to understand and use. The grain and oilseed markets use basis as a shorthand for trading. Sophisticated grain traders will offer cash corn for sale at "three off," meaning 3 cents under the nearby futures contract. Cash grain and oilseeds are thus usually bought and sold at prices that are relative to the futures price by some amount. The words "off" or "under" mean the cash price is below the futures price and "on" or "over" mean above the futures price. A trader that sold "two on" actually got a price for her grain that was 2 cents above the nearby futures price.

As the cash market is related to the futures via a negotiated price differential, the opportunity to hedge and manage the basis increases. Table 5–14 illustrates an example of a basis trade with a grain merchant. The merchant has spot wheat purchased at $3.50 per bushel and hedged at a price of $3.70 per bushel. The beginning basis (often called the buying basis) is 20 cents. Any trade to sell the wheat for less than 20 cents will result in a net gain for the merchant. If he negotiates to sell the wheat for "15 off" he will have a 5-cent net gain.

The merchant purchased the grain for $3.50 per bushel and had a net selling price of $3.55 per bushel, for a net gain of 5 cents per bushel—the change in the basis. The merchant assured himself a net gain by negotiating an ending or selling basis that was favorable relative to the beginning or buying basis.

The merchant in the example could have purchased the grain to begin with via a basis contract. The merchant could have offered to a wheat producer a basis contract with the following terms: transfer the title to the wheat now and keep the pricing rights to the wheat until November 30; up to November 30, any day's December futures price can be used and the wheat priced at 20 cents under the futures. In other words, the wheat

Table 5–14 Trade with a Net Gain

Cash	Futures	Basis
Nov 1, buys wheat at $3.50 per bushel	Sells Dec wheat at $3.70 per bushel	$0.20
Agrees to sell at "15 off" the December futures, currently at $3.40 per bushel, so the cash selling price is $3.25 per bushel		
Nov 20, sells wheat at $3.25 per bushel	Buys Dec wheat at $3.40 per bushel	$0.15
Net loss of $0.25	Net gain of $0.30	Change of $0.05

Net hedged selling price = $3.25 + $0.30 = $3.55

producer can select any day's December futures price and receive 20 cents less than that price for the wheat. If the merchant negotiated this basis contract with the wheat producer in October, then he has set the beginning basis and now knows what ending basis will be.

A basis contact must have as a minimum the following:

1. The futures contract, i.e., which delivery month, such as December

2. A negotiated differential relative to the futures price for the cash commodity to be priced, i.e., what is the basis ("three on," "two over," etc.)

3. An ending point date, i.e., up to 5 P.M. on November 30

4. Knowledge of when title to the commodity passes and how storage costs are handled

Basis contracts offer hedgers the opportunity to have better control over beginning and ending basis values and thus the net margins on trades. They are in widespread use in the grain and oilseed markets and reach sophisticated levels using provisions called call contracts.

Call Contracts

A call provision that is added to a basis contract specifies that the contract holder must call the broker of the contract provider to stipulate the day that the basis contract will be exercised. Otherwise, a **call contract** is just a simple basis contract. Proper use of call contracts allows certain hedgers to fix both sides of a trade.

A grain merchant enters into a call contract with a corn producer on September 1. The merchant will give the producer until November 30 to call and price the grain. The negotiated deal is this: the merchant will pay the producer the December corn futures price less 10 cents per bushel. The producer in turn releases the corn to the merchant on September 1, but has until November 30 to call the merchant's broker. The merchant's broker is informed of the deal and understands that when the producer calls, the broker will place a short futures position with December corn futures.

The merchant then enters into another call with a feed processor to provide corn on September 1. The processor gets the corn now but has until November 30 to price the corn. The deal the merchant makes with the processor is that the processor will pay the merchant the December corn futures price less 5 cents per bushel. The processor can call the merchant's broker any time up to November 30. The merchant's broker is informed of the deal and instructed to buy a December corn futures when the processor calls.

Both sides of the deal are struck and the merchant has made 5 cents per bushel on the deal. How? The merchant has fixed the basis on both sides of his deal—one at 10 cents and the other at 5 cents, thus the change is set for 5 cents and becomes the merchant's margin. Table 5–15 shows the deal with the processor starting the process and Table 5–16 indicates the same deal with the producer calling first. It doesn't matter which party calls first, nor does it matter what the futures price is. The merchant will still make the same 5-cent margin because the basis has been fixed for both the beginning and ending deals.

The call contract with the producer is called a *seller's call* because the action that the broker is instructed to take is a short futures position. The deal with the processor is called a *buyer's call* because the broker's action will be a buy in the futures market.

Table 5–15 Call Contract Example, Grain Merchant Brokerage Account with Processor Calling First

Cash	Futures	Basis
November 1, processor calls when the December corn futures is at $2.30 per bushel.		
Sells corn to processor at $2.30 less agreed upon basis of $0.05 per bushel, or $2.25	Buys December corn futures at $2.30 per bushel	$0.05
November 15, producer calls when the December corn futures is at $2.50 per bushel.		
Buys corn from producer at $2.50 less agreed upon basis of $0.10 per bushel, or $2.40	Sells December corn futures at $2.50 per bushel	$0.10
Loss of $0.15 per bushel	Gain of $0.20	Change of $0.05
Net gain of $0.05 ($0.20 – $0.15)		

Table 5–16 Call Contract Example, Grain Merchant Brokerage Account with Producer Calling First

Cash	Futures	Basis
November 1, producer calls when the December corn futures is at $2.30 per bushel.		
Buys corn from producer at $2.30 less agreed upon basis of $0.10 per bushel, or $2.20	Sells December corn futures at $2.30 per bushel	$0.10
November 15, processor calls when the December corn futures is at $2.50 per bushel		
Sells corn to processor at $2.50 less agreed upon basis of $0.05 per bushel, or $2.45	Buys December corn futures at $2.50 per bushel	$0.05
Gain of $0.25 per bushel	Loss of $0.20	Change of $0.05
Net gain of $0.05 ($0.25 – $0.20)		

Crush Hedges

Most soybeans go through a crush process that separates the raw beans into meal and oil. The old manufacturing process was an actual crushing machine that squeezed the oil out of the beans. Modern crush plants still crush the raw beans somewhat, but rely on solvents to extract the oil. The solvent is extracted and reused. Soybean processors buy raw soybeans, crush them, and sell the resulting meal and oil. The amount of meal and oil that a bushel of soybeans produces is called the crush yield. Crush yields are reported on a regular basis, but a general rule of thumb is that a 60 pound bushel of soybeans will produce 48 pounds of meal, 11 pounds of oil, and 1 pound of waste. The difference between the value of the meal and oil and the price of the soybeans is called the crush margin. When the crush margin is greater than the cost of crushing, processors will enter into a hedge called *putting on crush.* When the margin is less than the cost, the hedge is called a **reverse crush.**

The theory behind the processor crush hedge is that the difference between the value of a raw product and the value of the finished products obtained from the raw product should be equal to the cost of processing. If the margin is higher than the cost of processing, then the profits will bid up the price of the raw product as more processors try to buy the product to produce. As they produce more the final products will be worth less as more of the products reach the market place. This action will reduce the profit margins back down to the cost of processing. On the other hand, if the cost of processing is greater than the margin between the value of the raw product and the finished products, processors will stop producing as much. This will result in the price of the raw product falling as few processors bid in the market place and the value of finished products will increase as they become scarcer. This action will drive the margin back up to the cost of processing.

Soybean Crush

Soybeans are unique among futures contracts in that the raw product and both of the major items produced from the raw product—meal and oil—have futures contracts. Canola now has a meal futures contract, but not an oil contract, so a complete crush hedge is impossible. Table 5–17 is an example of a soybean crush hedge.

A soybean processor can buy November soybeans for $5.50 per bushel. The November soybean oil contract is trading at $0.10 per pound and the November meal contract is at $0.11 per pound. Using the crush yield of 48/11/1 the value of the meal is 48 × $0.11 = $5.28 in a bushel of soybeans. The oil value is 11 × $0.10 = $1.10 per bushel. The total value of the meal and oil in a bushel of soybeans is $6.38 per bushel. The beans are trading at $5.50 per bushel for a crush margin of $0.88 per bushel. Current crush costs are $0.50 per bushel. There is a net margin of $0.38 per bushel for the November futures. The processor will put on a crush hedge by buying soybeans and at the same time selling meal and oil futures.

By November 1 the processor lifted the hedge and made the net margin as the prices moved as the theory suggested. If the prices did not change by November 1, the processor could stand for delivery of the soybeans, process them and deliver against the futures and make the net margin. Processors that use **crush hedges** can effectively set their net crush margins in advance.

Reverse crushes are used when the net crush margin is less than the cost of crush. The hedge is a simultaneous selling of the soybeans and buying of meal and oil futures as illustrated in Table 5–18. The crush margin is zero and the net margin is a loss of the cost of crush. The reverse crush allows the processor to make back the cost of crush. This hedge will not yield a positive net crush margin, but does allow the processor to recover the cost

Table 5–17. Soybean Processor Putting on Crush Hedge

September 1,   November soybeans at $5.50 per bushel

November soybean oil at $0.10 per pound

November soybean meal at $0.11 per pound

Buy November soybeans at $5.50 per bushel

Sell November soybean oil at $0.10 per pound

Sell November soybean meal at $0.11 per pound

Value of November oil = 11 × $0.10 = $1.10 per bushel

Value of November meal = 48 × $0.11 = $5.28 per bushel

Total value of meal and oil = $6.38 per bushel

Value of November soybeans =$5.50 per bushel

November crush margin = $0.86 per bushel

Cost of crush = $0.50 bushel

Net crush margin = $0.38 per bushel

November 1,   **Sell November soybeans at $5.64 per bushel**

Buy November soybean oil at $0.10 per pound

Buy November soybean meal at $0.105 per pound

Net gain on November soybeans = $0.14 ($5.64 – $5.50) per bushel

Net gain on November soybean oil = $0.00 ($0.10 – $0.10) per bushel

Net gain on November soybean meal = $0.24 ($0.11 – $0.105 × 48)

Total net Gain = $0.38 per bushel

of crush. Soybean processing plants cannot easily be shut down and started up just because of a temporary loss. Without the reverse crush hedge the processor would have lost 50 cents per bushel, but with the hedge they broke even.

Using Futures in the Livestock Industry

Livestock products have a rich history in the futures industry. The forerunner to the Chicago Mercantile Exchange (CME) was the Chicago Butter and Egg Board in the late 1800s. Over the last 125 years, almost every type of livestock product from eggs to frozen beef has had a futures contract. No other futures contract enjoys such popularity (good and bad) as bacon—the infamous pork belly futures contract of television and movie lore as a way to speculate in the futures market.

When the CME first proposed live animal futures contracts in the early 1960s, they were greeted with a great deal of skepticism. Critics did not understand how a futures contract could exist on something that was not storable. Livestock products such as eggs, butter, and cheese can be stored and fit the cost of carry model for price differences between two futures months. Live animals cannot be stored for longer than a few days before the growth and aging process changes their form. Critics asked, "What would be the theoretical model for the price differences between two futures months?" The correct answer, one that critics

Table 5–18. Soybean Processor Reverse Crush Hedge

September 1, November soybeans at $5.90 per bushel
November soybean oil at $0.10 per pound
November soybean meal at $0.10 per pound

Sell November soybeans at $5.90 per bushel
Buy November soybean oil at $0.10 per pound
Buy November soybean meal at $0.10 per pound

Value of November oil = 11 × $010 = $1.10 per bushel
Value of November meal = 48 × $0.10 = $4.80 per bushel

Total value of meal and oil = $5.90 per bushel
Value of November soybeans =$5.90 per bushel
November crush margin = $0.00 per bushel
Cost of crush = $0.50 bushel
Net crush margin = ($0.50) per bushel

November 1, **Buy November soybeans at $5.75 per bushel**
Sell November soybean oil at $0.11 per pound
Sell November soybean meal at $0.105 per pound
Net gain on November soybeans = $0.15 ($5.90 – $5.75) per bushel
Net gain on November soybean oil = $0.11 ($0.11 – $0.10 × 11) per bushel
Net gain on November soybean meal = $0.24 ($0.11 – $0.105 × 48)

Total net gain = $0.50 per bushel (enough to cover the cost of crush)

didn't like, was "We don't know." In retrospect, the decision to offer live animal futures contracts became the watershed for other futures such as currencies, financial instruments, and indexes that followed a decade or more later. Futures had to be viewed with a wider angle lens than just storable commodities.

Livestock Basis

Live animal futures—hogs, feeder cattle, and live cattle—have a basis that does not follow the cost of carry model for price differences. To be sure, a few days of storage exists as both hogs and cattle can be held on a maintenance ration for a brief period of time, but not for any extended period. What should be the difference between today's local cash price and the next futures delivery contract? A small amount of storage (a few days at most), transportation (if not near a delivery point), and *expectations of future supply and demand* are the basis components of live animal futures. The storage component is small and almost nonexistent. If a transportation differential exists, it is a constant just like that for storable commodities. The major difference between a cash price and a futures price is the expectations of differing supply and demand conditions in the future versus today. Unfortunately, a useful model to predict the different expectations in supply and demand does not exist. In other words, we don't know—just as we didn't know in the 1960s when livestock futures were first proposed.

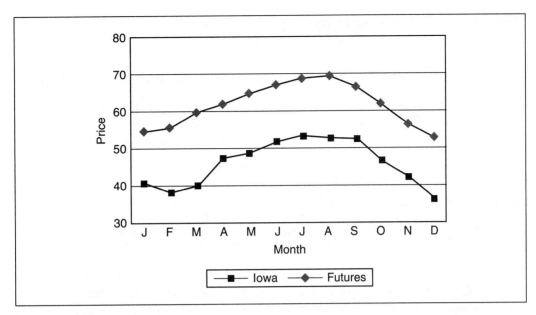

Figure 5–8   Iowa cash hog prices and lean hog futures prices, 2001

Hogs and cattle have seasonal and cyclic trends that can be useful to estimate price differences between time periods, but can serve only as a guide. Basis values generally narrow as the futures contract delivery month approaches the calendar date as expectations and reality get closer. However, basis is not as predictable for live animals as it is for storable commodities, and no satisfactory model exists that is a useful guide for estimating expected basis values. Traders resort to empirical data to get trends and patterns. Figure 5–8 shows that Iowa cash hog and futures prices do tend to trend together, thus the futures market can be useful as a price risk management tool.

Livestock products such as milk, butter, nonfat dry milk, and pork bellies (both fresh and frozen) more closely follow the cost of carry model because they can be stored for some period of time (fluid milk is the exception, but fluid milk can be held for 2 weeks so it does have a storage cost).

Production Hedges
Live animals such as hogs, feeder cattle, and live cattle go through a growing process that subjects them to the risk that prices will decline and they will be worth less than the cost of production when they are ready for market. The lean hog futures contract is for a slaughter-ready animal and likewise for the live cattle futures contract. The feeder cattle contract is for an animal that is ready to be put into a feeding operation to become a slaughter-ready animal (live cattle). The hog industry is composed of several stages; one stage includes producers who have a farrowing operation that produces only weaning piglets. Others have a finishing operation that buys the weaned animals and finishes them into slaughter-ready animals. Still others are farrow-to-finish operations. Regardless of the type of hog producer, since only one type of futures contract exists, only slaughter-ready animals can be hedged. Cattle production operations are conceptually similar. Ranches run the gamut from cow-calf operations that produce weaned calves to those that carry the weaned animals into a

Table 5–19   Hog Production Hedge

Cash	Futures
January 15, purchase 180 feeder pigs	Sell one April hog contract at $44/cwt.
March 30, sells 177 hogs at an average weight of 238 each for $36/cwt.	Buy one April hog contract at $38/cwt. Gain $6/cwt.
Total weight 42,126 pounds	Total income $2,400
Total income $15,165.36	
Total net hedged income $17,565.36	
Net hedged selling price = $41.70/cwt.	

Net hedged selling price = $36 + $6(40,000/42,126) = $36 + $5.70 = $41.70/cwt.

final finishing operation that sells slaughter-ready animals. However, cattle futures exist on feeder animals and live animals and therefore provide more risk management opportunities than hog futures contracts do for swine operators.

Hog Production Hedge

A hog operation that buys weaned pigs will put those animals on an intense feeding plan to get slaughter-ready animals. The feeder pigs are generally purchased near a weight of 60 pounds on average and will be fed for about 80 days on average. Table 5–19 shows a production hedge with a basis improvement. The producer estimates that the death loss will be two feeder pigs and that the slaughter-ready hogs will weigh approximately 230 pounds each at market time. When market time arrives, the producer will have had a death loss of three animals and the average weight was 238 pounds each. The producer received a net price for the hogs of $41.70 per hundredweight for 177 animals despite having a higher death loss than anticipated and selling somewhat heavier hogs. The producer was under-hedged from the beginning because the estimated pounds to be sold later were 178 (estimated death loss of two pigs) at 230 pounds each for a total weight at market estimate of 40,940 pounds. A hog futures contract is for 40,000 pounds. The producer actually sold 42,126 pounds, which made the hedge under-hedged by 2,126 pounds. Put another way, 2,126 pounds received the market price of $36 per hundredweight ($765.36) and 40,000 pounds received the benefit of the hedged price of $42 per hundredweight ($36 + $6), or $16,800 for a total of $17,565.36. Because the hog futures contracts, like the grain and oilseed contracts, are standardized, it is impossible for any producer to ever be perfectly hedged.

Cow-Calf Operation

A typical cow-calf operation involves a brood herd of cows that are artificially inseminated or bred with bulls. Calves are weaned normally in the fall and weigh between 350 and 600 pounds depending upon breed, feeding conditions, management, health, and weather. Many ranchers will either sell the animals at weaning or hold them to heavier weights. Ranchers will hold the animals to heavier weights if they have ample grazing available or want to sell them to feedlots. The feeder cattle futures contract calls for 50,000 pounds of heavy (greater than 800 pounds) feeder steers. Cow-calf producers will almost always have to place a *cross hedge*.

Table 5–20   Cow Calf Hedge

February 1, estimates 142 head to sell in fall	Sells two November feeder cattle contracts @ $75/cwt.
November 1, sells 144 head @ 605 pounds each (87,120 pounds) @ $75/cwt.	Buys two November feeder cattle contracts @ $68/cwt.
	Gain of $7/cwt.
Total sales = $65,340	Total gain = $7,000
(87,120 × $0.75)	(100,000 × $0.70)

Total net hedged income $72,340 ($65,340 + $7,000)

Total net hedged price = $83.03/cwt. ($72,340/87,120)

A cross hedge is a hedge whereby the cash and futures specifications do not match exactly. Almost all futures hedges are cross hedges, but the differences are much greater with feeder cattle. First and foremost, the futures contract calls for steers (males castrated before sexual maturity) while all cow-calf operations will have an approximately 50-50 split between steers and heifers (nonimpregnated females). Steers almost always carry a cash price premium. The second major difference is that feeder cattle futures are for heavy feeder steers (greater than 800 pounds) while most ranchers will sell their animals at weaning weights of 500 pounds on average, thus there is often a major mismatch of weights. Lighter weights carry a higher spot price premium per hundredweight. Despite these two major differences, light weaned mixed-sex feeder animals can be cross-hedged fairly well with the heavier feeder cattle contract because price trends will be similar.

Assume a rancher has 150 cows and estimates a calf crop of 145 animals (a 97 percent calving rate). The rancher estimates a death loss of three animals during the growing season and a net of 142 weaned animals to sell at an approximate weight of 600 pounds each. The total estimated weight of animals at market time will be 85,200 pounds (142 × 600). The futures contract calls for 50,000 pounds. The rancher will have to be under-hedged by 35,200 pounds with one futures contract, or over-hedged 14,800 pounds with two futures contracts. The rancher opts to be **over-hedged,** as Table 5–20 reveals.

The rancher had 87,120 pounds of animals to sell and was hedged with 100,000 pounds of futures for an over-hedged position of 12,880 pounds. The rancher got a net hedged price of $83.03 per hundredweight. The overall price of feeder cattle fell during the growing season for the rancher and the hedge earned an extra $7,000 in income for the rancher, pushing the net price up from the spot price of $75 per hundredweight to $83.03 per hundredweight.

Feedlot Hedge

Feedlots are manufacturing operations that process concentrated rations of grains and oilseeds and other nutrients through feeder cattle into beef that is sold to consumers ultimately as steaks, roasts, and other retail beef products. Feedlots (also called feed yards) put feeder cattle ranging in weight from approximately 600 pounds to 900 pounds on rations that will cause them to gain weight and change product form so that they are appealing to customers as retail beef products. Two major types of feedlots exist: custom feeders and retained owners. Custom feedlots solicit customers to put their own cattle in the feedlot

Table 5–21   Feedlot Hedge for Live Cattle and Corn Input

Cash		Futures	
Cattle	Corn	Cattle	Corn
Jan 1 Puts 500 feeder cattle on feed (current live cattle price @ $80/cwt.)	Has enough corn for first half of feeding period purchased @ $2.10/bushel	Sells 10 live cattle futures @ $84/cwt.	Buys 20 May corn futures @ $2.75/bushel
April 1	Buys 100,000 bushels of corn @ $3.00/bushel		Sells 20 May corn futures @ $3.65 bushels Gain of $0.90/bushel
July 1 Sells 500 fed (live) cattle @ $68/cwt. (average weight of 1,000 pounds each)		Buys 10 live cattle futures @ $72/cwt. Gain of $12/cwt.	

Net hedged selling price live cattle = $68 + $12 = $80/cwt.

Net hedged buying price corn input = $3.00 − $0.90 = $2.10/bushel

and pay a fixed fee for the feed, care, and management of the animals. The customer then sells the cattle to a packer. A retained-owner feedlot owns its own cattle, pays for the feed, care, and management of the operation, and sells its own cattle. Some feedlots do both. They put as many of their own cattle on feed as they want to and then solicit other custom feeding as a way to fill the yard to capacity.

A retained-owner feedlot has the price risk that while the animals are being fed, the price of live (fed) cattle will decrease. It will need a typical production hedge. Also, it will have to periodically buy grains and proteins to make the rations. It will be worried that grain and protein prices will increase. It will need to long hedge the inputs.

Table 5–21 illustrates a feedlot hedge for the live cattle (output) and the major feed ingredient, corn (input). The example shows no differences between the cash and futures quantities or any basis changes (that is, they are perfectly hedged) for both the output and input. This allows for a better view of the reason to hedge both output and inputs when possible. The feedlot could have sold live cattle in January for $80 per hundredweight but the cattle needed to be fed out until July. During that feeding period, live cattle prices fell to $68 per hundredweight; however, the hedge protected the feedlot and the end result was a net hedged selling price for the live cattle at $80 per hundredweight. A similar opportunity exists for one of the major feed inputs, corn. The feedlot could have purchased all of the corn needed in January, but had storage for only half the amount needed. The hedge protected the feedlot against the price of corn increasing, which it did by April 1 from $2.10 per bushel in January to $3.00. Yet the hedge lowered the net hedged buying price back to $2.10 per bushel. The process by which a manufacturer, such as a feedlot, hedges all available outputs and inputs is called *total hedging*.

Table 5–22   Dairy processor hedge

Cash	Futures
Feb. 1, forward sell milk @ $16.17/cwt. for delivery in 5 days (assume fluid milk price @ $10.17/cwt., processing costs @ $5/cwt., leaving a profit margin of $1/cwt.)	Buys fluid milk futures @ $12.50/cwt.
Feb 5, buys fluid milk to process and deliver on contract @ $11/cwt.	Sells fluid milk futures @ $13.33/cwt.
Loss of $0.83/cwt. ($11 – $10.17)	Gain of $0.83/cwt. ($13.33 – $12.50)

Processing Hedge

A milk processing plant buys milk from dairies. It pasteurizes and homogenizes the milk, packages it, and ships it out to wholesalers or retailers. A milk processor runs the price risk that between the time it buys the milk from dairies and sells it to a retailer the price will drop. It also forward sells to retailers for delivery at a later date and then has to buy the milk from dairies, and thus faces the risk that prices will increase. This type of a food processing company can be both a short and long hedger, depending upon which cash action it has performed first.

Assume a milk processing company gets a chance to forward sell fluid milk to a retail supermarket chain at a price of $2.10 per gallon ($16.17 per hundredweight equivalent at 7.7 gallons per hundredweight; processing costs are $5 per hundredweight, leaving a profit margin of $1 per hundredweight, assuming a cost of milk from dairies at $10.17 per hundredweight). The processor will deliver the milk in five days. The processor buys fluid milk futures at $12.50 per hundredweight. On day four they buy fluid milk from dairies at an average price of $11 per hundredweight and lift the hedge at $14.83 per hundredweight. Table 5–22 summarizes the transactions. No basis change or quantity differences are used so that the pure effect of price protection can be observed.

This example shows how a processor can forward sell product in advance and still protect profit margins. The processor still made a $1 per hundredweight profit even though the price they had to pay for the product went up $0.83 per hundredweight. Without hedging, the profit margin would have been reduced to only $0.17 per hundredweight. It is important to note that if prices had gone down, the cash purchase price would have been lower and the profit margin wider, but the futures transactions would have taken that away and the profit margin would still be $1 per hundredweight.

Processors and other businesses can fix the profit margin in advance and will receive close to that margin, plus or minus the basis change. Without hedging, the processor's margins will rise and fall with the spot market. This form of hedging is called *margin-based hedging* and is a tremendous risk management tool.

Despite not having a good theoretical model to predict basis in live animal futures, livestock futures can perform as a price risk management tool of considerable value. Most traders who regularly use livestock futures use historical basis charts that reveal the empirical values over time. The past is not the future, but the empirical basis does give traders a sense of basis levels during certain periods and what past trends show.

Contract	Size	Delivery Method
Chicago Board of Trade		
T Bonds	100,000	May, June, Sept, Dec
T Bonds	100,000	May, June, Sept, Dec
Chicago Mercantile Exchange		
T-Bills	1,000,000	May, June, Sept, Dec
1-Month LIBOR	3,000,000	All calendar months
Eurodollar	1,000,000	May, June, Sept, Dec
Canadian dollar	100,000 Canadian Dollars	May, June, Sept, Dec
Japanese yen	12,500,000 yen	May, June, Sept, Dec
Mexican peso	500,000	All calendar months

Figure 5–9.   Major interest rate and currency futures

Using Nonagricultural Futures in Agriculture

The most successful futures contract in terms of volume of trades for the last several years has been the Treasury Bond futures contract traded on the Chicago Board of Trade. Prior to the early 1970s only agricultural contracts or other commodities such as metals were traded on futures exchanges. Livestock contracts in the late 1960s exposed traders to dealing with nonstorable commodities, and as they learned to deal with more uncertainty, other ideas emerged. In the early 1970s foreign currencies and then later financial futures contracts were added to the trading pool. Figure 5–9 shows the major interest rate and foreign currency.

Most nonagricultural futures contracts involve large sums of money such as the T-Bond futures contract at $100,000 so they are not very useful to very small agribusinesses, but they can be very important to medium to larger operations. Agribusinesses use large amounts of credit and therefore have credit risks pertaining to changes in interest rates. Additionally, any agribusiness that deals overseas will potentially have a foreign exchange risk as financial payments get converted to and from dollars and other currencies. Interest rate risk and currency risk will be discussed in this section and basic hedging with futures will be outlined. More complex risk management strategies will be developed in other chapters using nonagricultural futures as they relate to overall operational risks in agriculture.

Interest Rate Risk

Futures contracts exist on several interest rate instruments as shown in Figure 5–9. These contracts cover short-term (T-Bills), intermediate (T-Notes), and long-term (T-Bonds) interest rates as well as numerous others such as LIBOR and municipal bonds. Most of these contracts do not directly relate to any agribusiness or producer, yet they can be used to mitigate certain risks with interest rates via cross hedging. T-Bills are 90-day short-term riskless interest rates. U.S. Treasuries—bills, notes, and bonds—carry zero risk of default and thus are called riskless. If the U.S. Treasury fails, so the theory goes, all is lost anyway. Conversely, a bond issued by a private company will have a higher interest rate than a T-Bond because it will carry some risk of default because companies can fail (Enron and

Montgomery Ward are recent examples). From a private investment standpoint, the difference between private debt instruments and U.S. Treasury interest rates is important, but from a hedging standpoint all that matters is that the prices of U.S. Treasuries move as the general market for debt instruments moves. Since the U.S. Treasuries represent a riskless rate, they form the base for all interest rates. As the base value moves, so do all other rates. This is the reason that the U.S. Treasuries have become so popular as a way to protect against all interest rate movements.

Two other futures contracts are important: the Eurodollar and one month LIBOR. The Eurodollar futures represents U.S. dollars that are deposited in banks outside the U.S. and represent short-term interest rates on corporate accounts. The London Interbank Offer Rate (LIBOR) is a very short-term rate that is a benchmark for corporate financing and is tied to a variable rate.

Interest Rate Hedging

Agricultural firms that use debt financing can either get a fixed rate loan or a variable rate loan. Fixed rate loans do not have any risk that the interest rate will change during the loan period. Variable rate loans are subject to change based on a fixed schedule, often quarterly or annually. Variable rate loans have the risk that rates will move against the borrower, i.e., go up.

Variable rate loans usually carry a lower initial rate relative to a fixed rate loan. The lower initial rate reflects a lower risk to the loan originator. A variable rate loan will be adjusted as interest rates move up or down, thus the financial institution that made the loan does not bear the risk of the interest rate change—the borrower does. A fixed rate loan has to have a high enough rate to compensate for the risk of an adverse interest rate movement for the financial institution. The financial institution bears the risk of rate changes with a fixed rate loan, not the borrower.

Assume an agribusiness firm borrows $1,000,000 for six months to meet some unusual cash flow needs. The rate is tied to the T-Bill rate and can be adjusted every quarter. The formula for the interest rate on the loan is the current T-Bill rate plus 5 percent. The current rate on the day the loan was completed was the T-Bill rate (3.5 percent plus 5 percent = 8.5 percent). In three months the rate will adjust once again by the formula. The firm has the loan rate fixed at 8.5 percent for the first three months but has interest rate risk for the last three months. If T-Bill rates go up, the loan's rate will also go up and the borrower will have to pay more. For the firm to protect themselves, they need something that will increase in value when interest rates increase. As interest rates increase, the price of interest rate instruments falls, so the borrower needs to hedge short with T-Bill futures as illustrated in Table 5–23. The firm had to pay 1 percent more on the loan for the last three months of the maturity, for an extra out-of-pocket expense of $2,500. The hedge protected against interest rates going up and brought in an additional $3,000 to more than offset the extra cost of the loan. The loan's effective interest rate was lower.

Interest rate hedging is trickier than commodity futures hedging because time has to be considered in valuing interest rate instruments and because of the inverse nature between the price of the instrument and the movement in interest rates. Later chapters will provide more detailed risk management examples with hedging and interest rates and the importance of measuring dollar equivalency between the cash exposure and the hedge.

Foreign Currency Risk

The dollar changes in value relative to many other currencies on a continuous basis, just like any other commodity changes in value depending on supply and demand factors. For the last

Table 5–23   Interest Rate Hedge

Cash	Futures
Jan 1, variable rate loan $1,000,000 for six months subject to change each quarter by the formula T-Bill Rate + 5%	Sell T-Bill futures at a price of $0.977 face value of contract ($1,000,000 × 0.977 = $977,000)
Loan rate for first 3 months = T-Bill rate (3.5%) + 5% = 8.5%	
April 1, loan re-prices at T-Bill rate (4.5%) plus 5% = 9.5%	Buy T-Bill futures at a price of $0.974 ($1,000,000 × 0.974 = $974,000)
Extra cost of loan (1% × $1,000,000) × 3 months ($10,000) × 3 months $2,500	Gain of $3,000 ($977,000 − $974,000)
Net gain of $500 ($3,000 − $2,500)	

30 years, most of the world's currencies have been openly traded instead of being quasi-fixed as was the case while the old Bretton Woods Agreement was in effect from the late 1940s until the early 1970s. The U.S. dollar is said to be strong when it takes fewer dollars to buy a foreign currency and is weak when it takes more. If it currently takes 65 cents U.S. to buy $1 Canadian and later in the week it takes 68 cents U.S. to buy a $1 Canadian, the U.S. dollar is said to be weaker relative to the Canadian dollar.

This is counterintuitive in a way. The higher the price of the U.S. dollar relative to buying Canadian dollars, the weaker the U.S. dollar, and vice versa. When other commodities such as corn increase in price, the commodity has also increased in value. Not so with currencies. Currencies are always priced relative to another currency, so as one increases in value, the other must decrease in value. Because they are priced relative to each other, an increase in the price of one must imply that fewer units of the other currency can be exchanged.

Exchange rates reflect the price of one unit of currency necessary to buy one unit of another currency. In the United States, if a person wanted to buy Canadian dollars, the exchange rate would be quoted as the price of one unit ($1 U.S.) to purchase one unit of Canadian currency ($1 Canadian), or as 0.65, meaning it takes 0.65 × $1 U.S. to buy $1 Canadian, or 65 cents U.S. In Canada, they will use the reciprocal (1 divided by .65 = 1.54). A Canadian would have to put up $1.54 Canadian to buy $1 U.S. Therefore, it is important to know how the currencies are stated relative to each other. Usually, within a country the exchange rate is expressed as how much of that country's currency is needed to purchase another country's currency. In Canada the U.S./Canadian exchange rate would be listed as 1.54, while in the U.S. it would be listed as 0.65—the same concept, just the reciprocal. This means there are two exchange rate numbers for each pair of currencies, as the example with Canada and the U.S. shows. When using foreign currency it is important to know and understand how the currencies are priced relative to each other so that proper adjustments and exchanges can be made.

The price risk with foreign currencies necessitates a flow and exchange between two or more currencies. Domestic agribusinesses that deal only in U.S. dollars need not worry about foreign exchange risk. Foreign exchange risk is added to the already-existing bundle of risks that firms must face when international financial dealings occur. Some companies avoid the risk of foreign exchange by demanding payment in U.S. dollars. Everything is priced f.o.b. U.S. dollars. Almost all U.S. cotton that moves to Mexico is priced by cotton merchants in "U.S. dollars f.o.b. border," meaning the Mexican merchant will pick up the cotton at a point on the U.S./Mexico border and the transaction will be in U.S. dollars. The U.S. merchants have shifted the risk of foreign exchange to the Mexican merchants. The Mexican merchants have all the risk of foreign exchange.

Foreign Exchange Hedging

If a U.S. cotton merchant wanted to win some new customers among Mexican cotton mills, one way to do it would be to shift the risk of foreign exchange from the Mexican business to the U.S. business. What if the U.S. cotton dealer priced the cotton to a Mexican cotton mill as "Pesos f.o.b. border"? The Mexican merchant does not have any foreign exchange risk—they will pay in their own currency at the border and take the cotton. The U.S. cotton dealer will now have pesos, and he will have to convert them to U.S. dollars. All of the risk of foreign exchange is now in the hands of the U.S. cotton dealer. If the U.S. cotton dealer can handle the risk of foreign exchange, then he will have a competitive advantage over other U.S. merchants that force the foreign exchange risk on Mexican merchants.

Table 5–24 presents an example of a U.S. cotton merchant's deal with a Mexican cotton mill. The U.S. cotton dealer forward sells cotton to a Mexican cotton mill for delivery in two weeks "Pesos f.o.b. border." The U.S. merchant would have lost 5.5 cents per pound on the

Table 5–24   Foreign Exchange Hedge

Cash	Futures
Jan 1, sells 100,000 pounds of cotton to Mexican cotton mill at 5.45 pesos per pound Current exchange rate = $0.11/peso U.S. dollar price per pound = $0.11 × 5.45 = $0.60 Contract value = 545,000 pesos Contract value = $60,000	Sells one March Mexican peso futures at $0.115 (500,000 pesos per contract) Contract value = $57,500
Jan 15, cotton delivered to border crossing. U.S. merchant receives 545,000 pesos Current exchange rate = $0.10/peso Contract value = 545,000 pesos Contract value = $54,500 Loss of $5,500	Buys one March Mexican peso futures at $0.105 Contract value = $52,500 Gain of $5,000

Net hedged selling price = $54,500 + $5,000 = $59,500/100,000 = $0.595 per pound

exchange rate risk had they not hedged. The hedge protected the merchant almost completely because the net hedged selling price was $0.595 per pound versus $0.60 per pound when the deal was struck. The advantage to the U.S. merchant is that they need not fear foreign exchange risk because proper hedging can counterbalance the risk. International markets are available to businesses that know how to handle the risk of foreign exchange.

Synopsis

Futures contracts allow agricultural businesses and producers to manage the risk of price change. The contracts are simple to use—hedges are either long to protect against increasing prices or short to protect against decreasing prices. A combination of simultaneous buys and sells are used by soybean processors when they protect crush margins. Farmers and ranches do not have to be at the mercy of major market price swings if they hedge with futures. The downside, of course, is if prices move in favor of the hedger, then the hedge will take that gain away. Price risk management with futures has a major cost— it protects against adverse price movements at the expense of taking away gains when favorable moves occur. To compensate for this major drawback, futures hedgers use basis trades and speculate on when is the best time to place the hedge, called **selective hedging.** Selective hedging and basis trading are important concepts in price risk management and will be dealt with in more detail in later chapters as sophisticated risk management strategies are developed. Nonetheless it is critical to understand the fundamentals of futures hedging extremely well. The basic futures hedge is 90 percent of all price risk management. To know futures hedging is to have the most importance piece of price risk management mastered.

CHAPTER 5—QUESTIONS

1. A cotton farmer plants in the spring when local cotton prices are at $0.60 per pound. The farmer properly hedges. Assume that the amount the farmer sells in the cash market at harvest and the size of the futures position exactly match. The cotton farmer sells in the fall after harvest at a local price of $0.50 per pound. The farmer had a basis improvement of 2 cents per pound. What would be the farmer's net hedged selling price?

2. Soybean crush hedges are more certain to protect crush margins than reverse crushes. Why?

3. A rancher estimates that he will have 75,000 pounds of feeder cattle to sell at a later date. He must decide to over-hedge with two futures contracts (50,000 pounds each), or under-hedge with only one. Either way they are over- or under-hedged by the same amount (25,000 pounds). Which way should they hedge? Why?

4. A grain merchant buys cash corn for a price of $2.00 per bushel and properly hedges at $2.20 per bushel with a December corn futures contract. The merchant later sells the cash corn for $2.30 per bushel and lifts the hedge at $2.40 per bushel. What is the merchant's net hedged selling price? Assume quantities in the cash and futures are identical.

5. A grain merchant buys 9,000 bushels of cash wheat for a price of $3.50 per bushel and properly hedges at $3.70 per bushel with two July wheat futures contracts (5,000 bushels each). The merchant later sells the wheat in the cash market for $3.70

per bushel and lifts the hedge at $3.80 per bushel. What is the merchant's net hedged selling price?

6. A candy manufacturer forward sells candy to a retailer for delivery in two weeks. The candy uses a great deal of whole milk. The manufacturer prices the candy to the retailer assuming a fluid milk price of $10 per hundredweight. The manufacturer will buy the milk from a local dairy two days before delivery and manufacture the candy and then deliver it by the two-week deadline. What kind of hedge will the manufacturer place? Why?

7. A large agribusiness firm has a commitment from a bank for a loan of $10 million in two months. The bank will give the firm a fixed rate loan, but will not commit to what the rate will be until the actual day the loan is processed in two months. Does the agribusiness firm have an interest rate risk? If so, how would they handle the risk? If not, why?

8. A cattle trader buys feeder cattle from a rancher for $80 per hundredweight and properly hedges them at $82 per hundredweight. The trader can sell the cattle the next day to a feedlot at a basis of $2.30 per hundredweight. Should the trader take the feedlot deal? Why?

Fundamentals of Options Hedging

KEY TERMS

option	price insurance	covered writing
double derivative	at the money	naked writing
call	in the money	target price
put	time value	synthetic futures hedge
premium	intrinsic value	second best
strike price	writing	

The objective of this chapter is to introduce one of the most powerful price risk management tools—**options.**

OVERVIEW

Options have existed for thousands of years. They have been known throughout history as "rights to buy" or "privileges." In the early 1900s privileges incurred a nasty reputation as highly speculative instruments with a high degree of default. The U.S. Congress banned privileges in agricultural and natural resource commodities through the 1936 Commodity Exchange Act. Options continued to be popular in real estate and in the securities business but never evolved in agricultural commodities after 1936 due to the ban. However, during the early 1970s several traders found a catch in the 1936 law that exempted "international" commodities. An active off-exchange market for options on rubber, cocoa, coffee, and some metals was created. It crashed in the late 1970s amid similar circumstances in the 1920s and 1930s—defaulted contracts. Meanwhile, options on equity stocks flourished and the Chicago Board Options Exchange became a major player in exchange-traded stock options when it opened for business in 1973. Real estate options and options on the actuals, especially on metals, continued to be popular. However, because of the hullabaloo over the failed return of options in the 1970s, the Commodity Futures Trading Commission (CFTC) banned all options on commodities, even the ones used successfully by physical traders. It took several court cases, which were decided by the U.S. Supreme Court, to restore option trading on the actuals. The CFTC opened pilot programs on options on futures in 1980 and by 1982 most exchanges had some option contracts available for trade. By the mid-1980s most futures contracts had options contracts. All major agricultural futures contracts now have options contracts and additionally there are numerous option contracts being traded off-exchange on most major agricultural commodities.

Futures contracts derive their value from the underlying cash commodity. An option on a futures contract derives its value from the underlying futures contract, which in turn derives its value from the underlying cash, thus the name **double derivative.** Double derivatives are also known as two-step derivatives. Options on the physicals such as equity stocks, real estate, and actual commodities are, however, only one-step derivatives because their value is directly derived from the underlying commodity. Only options on futures contracts are double derivatives.

These double derivatives have emerged as one of the foremost tools in managing the risk of price change in agriculture. Options are much more versatile than futures contracts and can be structured to provide price risk management through a wide spectrum of price levels.

Option Basics

The buyer of an option is buying the right, but not the obligation, to buy (**call**) or sell (**put**) something at some point in the future. The seller of the option has an obligation to perform if the buyer *exercises* the option. If the buyer decides to *let the option expire* then the seller's obligation is dissolved. The seller will demand compensation for taking on the responsibility to perform. That compensation is called the **premium.** The price that the commodity will exchange at should the option be exercised is called the **strike price.**

Option sellers are also called writers, underwriters, grantors, and the ubiquitous short. Option buyers are simply buyers or long. Options on futures are just like futures in that they are strong contracts and can be retraded. An initial buyer of an option can do three things: exercise, let expire, or retrade by selling. An initial seller of an option can actively do only one thing—retrade by buying. If the initial seller does not retrade the option, he will have to passively wait for the buyer to either exercise or let the option expire.

The Insurance Concept

The basic premise of insurance is that the owner of a policy has purchased the right but not the obligation to use the benefits of the coverage. The underwriter of the policy has agreed to be obligated to provide the coverage in the event the owner wants the coverage. The buyer pays a premium to the underwriter who agrees to provide a certain level of coverage. With a typical homeowner policy for fire coverage the seller would be the insurance company while the buyer would be the homeowner, the owner of the policy. The buyer pays the insurance company a premium for a certain level of coverage (strike price). If no fire occurs, the buyer would let the option expire and the insurance company would keep the premium. The buyer would then buy another policy for the next year. If a fire occurs, the buyer would probably exercise the right to the coverage (but is not required to). The insurance company would then be obligated to perform according to the contract terms.

Because options are so similar to insurance, the terms and concepts are almost interchangeable. Options on futures contracts can correctly be called **price insurance.** The buyer buys a certain level of price protection (strike price) and pays a premium for that protection. If the price protection is needed, the buyer exercises the option and if the protection is not needed, the option is allowed to expire. Futures contracts cannot provide this insurance-like protection. Furthermore, options and insurance are similar in that insurance companies are regulated by state insurance departments and options are regulated by exchanges and the CFTC.

Premium and Strike Prices

Options on futures are traded or offered on exchanges at various strike prices at fixed intervals above and below the futures price with correspondingly different premiums. Each day the exchanges always offer three strike prices above the current futures price, one near the current price and three below. As a minimum there will always be seven strike prices

	Strike Price	Premiums	
		Puts	Calls
July 1			
	2.70	33	3
	2.60	27	6
December	2.50	21	9
Corn futures $2.40	2.40	15	15
Price	2.30	9	21
	2.20	6	27
	2.10	3	33
July 2			
	2.80	33	3
	2.70	27	6
	2.60	21	9
December	2.50	15	15
Corn futures $2.50	2.40	9	21
Price	2.30	6	27
	2.20	3	33
	2.10	1	42

Figure 6–1   Option strike prices and premium values for December corn futures contract

available for each delivery month. As the market price of the underlying futures contract changes, additional strike prices will be added. Toward the expiration date, literally dozens of strike prices will be available for trading. Figure 6–1 illustrates the concept with a December corn futures contract.

The July 1 relationships show three strike prices above the December corn futures price of $2.40 per bushel and three below. On July 2 the December corn futures price had moved up to $2.50 per bushel. A new $2.80 strike price had to be added to get three above the market and three below. Also, the $2.10 strike price is still active, providing four strike prices below the market. As the December corn futures price changes, the number of potential strike prices will increase.

In Figure 6–1 the strike price that is the same as the underlying futures price ($2.40 per bushel on July 1 and $2.50 per bushel on July 2) is called **at the money.** For puts (calls), the strike prices that are above (below) the at the money are said to be **in the money.** *Out-of-the-money* strike prices for puts (calls) are below (above) the at-the-money strike price. Some traders refer to in-the-money option strike prices as *on the money* and out of the money as *off the money.*

An at-the-money strike price is composed of only **time value.** On July 1, an at-the-money strike price on a call option on December corn futures is trading at a price of 15 cents per bushel. The buyer of the call option pays a premium of 15 cents per bushel to the writer for the right, but not the obligation, to have a long position in December corn at the strike price of $2.40 per bushel when the underlying December corn futures is trading at $2.40 per bushel. If buyers exercise the option immediately, they would have a long position in December corn at $2.40 per bushel for which they paid 15 cents per bushel. They could bypass the option and simply buy a December corn futures at $2.40 per bushel and not have to pay the 15 cents premium. Clearly the option buyer in this case has to have some value for the 15 cents and that value is time. The buyer is speculating that over time the December corn futures price will increase and having the right to buy it at $2.40 per bushel will be worth more than 15 cents per bushel.

The length of time to expiration of the option is directly related to the time value—the longer the time to expiration, the higher the premium. Obviously an option buyer will want the premium as low as possible, thus it is the seller of the option that plays the major role in premium values. If time is short to expiration, the probability that the price will move such that the buyer will exercise the option is lower than if the time frame is longer. Given a sufficiently long enough time horizon, prices can move very dramatically. The probability that a large meteor will hit the earth tomorrow is very small; however, if the time horizon is long enough, the probability will be one—it will happen with certainty. *Time value for options is really the probability of a price move.* Longer time periods have higher probabilities that prices will move, and vice versa.

Option premiums also have **intrinsic value.** Intrinsic value can be positive or negative, but only positive values are counted. All in-the-money options have positive intrinsic value. All out-of-the-money options have negative intrinsic value. In Figure 6–1 on July 1, the $2.20 strike price for a call has a premium value of 27 cents per bushel. The underlying futures contract is trading at $2.40 per bushel. If the buyer of the call exercised the call option, he would have a long position in December corn futures at the strike price of $2.20 per bushel. He could then sell the futures contract at the market price of $2.40 and have a net positive difference of 20 cents per bushel. The seller of the $2.20 per bushel strike price when the market is $2.40 per bushel will demand at least the 20-cent difference between the strike price and the market price for a premium. The difference between the strike price and the underlying market price is labeled intrinsic value. The premium for the $2.20 strike price is 27 cents. This 27 cents represents 20 cents of positive intrinsic value and the remainder, 7 cents, is time value.

On the other hand, the $2.70 strike price on July 1 for the call option is trading for 3 cents premium. If the buyer exercises the option, they will have a long position in the futures at $2.70 per bushel and have to sell at the current market price of $2.40 per bushel for a 30-cent loss. This 30-cent loss is a negative intrinsic value and consequently not counted. This out-of-the-money option will have value only if the market price moves so it only has time value (3 cents).

Positive intrinsic value is merely a bookkeeping concept. Sellers demand as a minimum the built-in profit (intrinsic value) of an in-the-money option. All that matters in the option premium is the time value. Every seller and buyer will have a different set of values for what they think will occur over time with the price. Since time value is really just the probability of a price move it becomes a function of how variable prices are during a given period and the length of the time periods. Therefore, a good rule of thumb for assessing time value of options is to *look at the variability of prices and the length of time to maturity. Other things being equal, the more variable the prices, the higher the premium; and the longer the time period, the higher the premium, and vice versa.* Option premiums are very important and over the last several years pricing models have been developed to address the trade-off between price variability and time.

Options Pricing: The Premium

The premium that a writer receives for an option and thus the amount that the buyer pays is very simply the price at which the options contract trades. Like every other good or service in an economy, the price at which the option trades is determined by supply and demand—the interaction of the interests of buyers (and their willingness to take on the risk associated with writing either a put or a call relative to the compensation they will receive to accept the additional risk) and sellers (and the cost of shifting the risk of adverse price movements relative to the benefits of risk avoidance). However, there is a third category of market participants, arbitragers, that play a role in options markets just like they do in every other market.

Before we consider the role of arbitragers, think about the major players in the markets—the buyers and the sellers. The buyers observe a price (the premium) in the market. For any given trading day, the strike prices are fixed, the length to maturity is fixed, market interest rates are generally steady (they will not usually change substantially in one day), and even though the price of the underlying futures contract is changing in the futures market, that price usually does not change within a range as large as it might over a week, over a month, or over six months. But, as all of these variables change, so does the premium—the trading price of the contract. All of these factors affect the supply of and demand for options contracts. Any one buyer or any one seller has relatively little market power—power sufficient to change price—especially if the buyer is seeking to hedge this year's crop of corn or calves or this month's wheat needs for milling.

Consider a single farmer selling his grain cartload of soybeans at the local grain elevator. That farmer has little power by himself to negotiate with the elevator manager to change his posted daily soybean price. The elevator manager is likely to believe that the next few farmers are already unloading their soybeans and taking the posted price even while our negotiator is laying out his story. The buyer (the elevator manager) himself has little power because he must retrade the soybeans, and market forces up the channel from him limit his power to pay more. He can, of course, pay the individual seller more but that will only cut into his profit. Similarly, the farmer could accept a price lower than the posted price, but that won't help him sell more.

In a very similar fashion, any one options buyer and any one options writer face a market price determined by market forces just like those faced by participants in the local grain market (Figure 6–2). If the premium increases, options sellers want to write more contracts (increase quantity supplied), if all other factors (underlying futures price, time to maturity, interest rates, and price volatility) are held constant in the short run. Conversely, options buyers would buy less as the price increases. Quickly, equilibrium—the trading price or premium—will be reached, or very few trades will take place. What would happen if price (premium) were too high to clear this market? Buyers would attempt to buy more than sellers would be willing to sell. The mechanism is the same for options contracts as it is for futures contracts or for the actual commodity. But how can a buyer or a seller really determine whether a given price, market-clearing or not, will meet his needs, especially if that buyer or seller finds himself in the position of being a price taker?

The Risk and Mitigation Profile process used in previous chapters is the best approach for a hedger to determine whether any particular options contract, along with its associated strike price and premium, meets his needs. Because the seller is assuming additional risk rather than offsetting it through a hedge, the RMP process is not the correct approach, but a similar assessment of the risk/return trade-off is necessary.

Arbitragers are aware of and participate in every market, from corn to cotton to sugar futures to soybean options to toys at garage sales and flea markets. Arbitrages keep markets

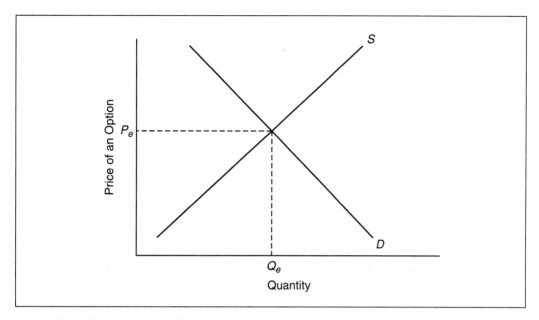

Figure 6–2   Supply, demand, and equilibrium price and quantity
of options contracts

honest and liquid. An arbitrager is always seeking an opportunity to exploit market imperfections. Suppose Elevator A was offering to sell corn at $3.25 per bushel and Elevator B (over 100 miles away from Elevator A with at least 25 other country elevators between) was offering to buy at $3.40 per bushel. Arbitrager C, who had stopped by Elevator A to check today's prices and have a cup of coffee, knows that regardless of the long-run volatility of corn and diesel prices or of interest rates and options premiums he can load his truck at $3.25 per bushel and deliver to Elevator B and return home to Elevator A for another load for $0.08 per bushel. Before the morning is over, he can net $0.07 per bushel. He won't be able to do this too often before Elevator A raise its price or Elevator B lowers its price, or both.

Options traders arbitrage in a fashion that is more mathematically sophisticated than the one used by our grain hauler, but the process is directly analogous. The primary tool that they use to determine if there are arbitrage opportunities (no matter how brief or how minute) is called the Black-Scholes Options Pricing Model (BSOPM). Although it has been modified somewhat over the years and subjected to a great deal of criticism, research, and refinement attempts, the BSOPM used in the industry today is remarkably the same as the model originally developed by Fischer Black and Myron Scholes in their seminal article published by the *Journal of Political Economy* in 1973.[1]

Before we investigate the BSOPM, it is helpful and reasonably easy to determine the maximum and minimum price of an option. Assume the premium to be zero for a moment. What is the maximum price a rational buyer would be willing to pay? Since the call buyer

---

[1] Black, F. & Scholes, M. (1973) "The pricing of options and corporate liabilities," *Journal of Political Economy, 81* (May/June), 637–659. Scholes was awarded the Nobel Prize in Economics in 1997 for this work; Black died in 1995.

$$CP = P\,N(d_1) - S\,e^{-rt}\,N(d_2)$$

where: $\quad d_1 = \dfrac{\ln(P/S) + (r + \sigma^2/2)^t}{\sigma\sqrt{t}}$

and $\qquad d_2 = d_1 - \sigma\sqrt{t}$

$\qquad$ CP = call premium

$\qquad$ P = current security price

$\qquad$ S = option strike price

$\qquad$ e = base of natural logarithms

$\qquad$ r = riskless interest rate

$\qquad$ t = time until option expiration

$\qquad$ σ = standard deviation of returns of the underlying security

$\qquad$ $N(\bullet)$ = cumulative standard normal distribution functions evaluated at $\bullet$

$\qquad$ $\ln$ = natural logarithm

Express r, t, and σ on the same temporal basis. Use days, weeks, months, exclusively; do not mix time periods.

Figure 6–3   The Black-Scholes Options Pricing Model

receives the right to buy a futures contract, the maximum price she would be willing to pay is the price of the futures contract. Conversely, the minimum she could ever expect to pay is the intrinsic value, which is futures price minus strike price but always a non-negative amount. For a put buyer, what are the maximum and minimum prices he would be willing to pay or expect to pay? If futures prices fell to zero, the maximum value of his option contract is the strike price, so the maximum price of an option would be its strike price. Its minimum price, like the minimum price of a call, would be its intrinsic value. Of course, the actual price of an option, either a call or a put, will fall somewhere between the theoretical maximum and minimum. The exact amount depends on many factors, factors that are included in the BSOPM. Figure 6–3 describes the BSOPM.

The mathematical equation looks like it is difficult to compute and to interpret, but modern spreadsheets, especially those that can be programmed onto handheld computers, reduce the mathematical complexity into a very accessible, readily available, and very useful tool. For example, check out www.fis-group.com/finmodls.htm for a variety of BSOPM applications.

The BSOPM was developed for European options on stocks—corporate equity securities, not commodity futures. This influence is clearly seen in Figure 6–4. But, the model can be adapted easily to options on futures contracts, and is in practice. Pit traders do not use their spreadsheets to recompute the mathematical model for small changes in every variable, just like equities traders do not recompute the discounted present value model of future cash flows for every stock market trade. But, they fully and intimately know how the model reacts to a change in any of the variables. Their success as pit traders, as arbitragers, depends on it.

Factor	Measure	Computation	Values
Premium change relative to futures price change	Δ	$\Delta = \Delta CP/\Delta P = \partial CP/\partial P$	$0 < \Delta < 1$ $\Delta \to 1$ Deep in the money $\Delta \to 0$ Deep out of the money $\Delta \to 1$ At the money
Rate of change of Δ	Γ	$\Gamma = \partial^2 CP/\partial P^2$	$0 < \Gamma < \infty$ $\Delta \to 0$ Deep in or out of the money
Time	Θ	$\Theta = \partial P/\partial t$	$0 < \Theta <$ option value, decays as option approaches expiration date
Volatility	Λ	$\Lambda = \partial P/\partial \sigma$	$0 < \Lambda <$ option value, decays as option approaches expiration date
Interest rate	ρ	$\Lambda = \partial P/\partial r$	$0 < \rho < \infty$

Figure 6–4   Options premium factors

The BSOPM was built and remains dependent upon several assumptions:

1. All relevant markets (for options, futures, and underlying actuals) are efficient.

2. Interest rates are constant and known.

3. Returns are lognormally distributed random variables.

4. No commissions are paid.

5. The option can only be exercised on expiration.

The last assumption forces the model to specifically address European options rather than the retradeable, strong form American options used in the agricultural commodities markets. Current BSOPM forms adjust for this relaxation of the original assumption; exercise before expiration date shortens it. Commissions are not important to a pit trader and are simply an additional charge to be entered into a hedger's RMP. Interest rates are constant (or nearly so) on any given day.

The BSOPM yields a theoretical premium that may or may not be exactly the same as the market premium (the premium being quoted currently when buyers and sellers enter the market). Market imperfections or differences between reality and assumption cause disparity between theoretical and market premiums. Especially relevant are:

1. The true underlying distribution of prices is not exactly lognormally distributed.

2. Borrowing and lending interest rates are different.

3. Volatility expectations differ among market participants.

4. The market is less liquid than it usually is or is theoretically expected to be.

Regardless of relatively minor theoretical/market inconsistencies, all traders use the BSOPM or one of its close variants. Consequently, the BSOPM remains the cornerstone of modern options premium determinations in the marketplace.

Astute arbitragers assess differences between theoretical and market premiums, as well as relationships between call and put premiums on the same maturity and relationships among premiums across maturities. But more important than the assessment phase, astute arbitragers exploit those differences and inconsistencies. Exploiting temporary differences forces markets relentlessly toward a fundamental equilibrium.

Options Pricing: The Greek Values

The BSOPM is based upon several major variables. From Figure 6–3, we see P (the price of the underlying futures contract), S (the strike price), r (the interest rate), t (length of time to option maturity), and σ (a measure of the volatility of futures contract prices). If all other things are held constant, as the futures price P increases, CP (the premium) would be expected to increase. That relationship between premium and futures price is referred to as delta (Δ), as shown in Figure 6–4. Delta is computed as the rate of change of the premium relative to the futures price, $\Delta CP/\Delta P$ or more correctly, $\partial CP/\partial P$ in a continuous mathematical model. Gamma (Γ) measures the rapidity of the change in premium relative to the futures price, thus it is the second derivative, $\Gamma = \partial^2 CP/\partial P^2$. Theta ($\Theta$) measures how quickly the time value of an option decays (see Figure 6–5), and is computed as $\Theta = \partial CP/\partial t$. The responsiveness of premium to volatility in futures prices is captured by lambda (Λ), thus $\Lambda = \partial CP/\partial\sigma$. And finally, rho ($\rho$) reflects the responsiveness of premium to the interest rate; thus $\rho = \partial P/\partial r$.

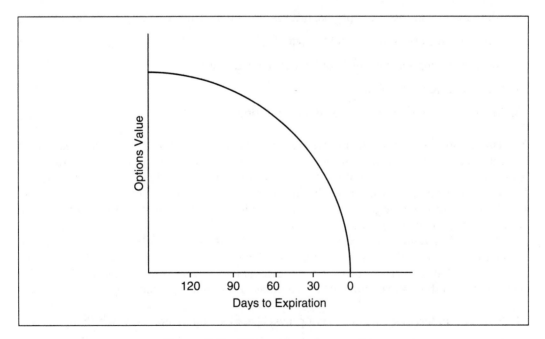

Figure 6–5   Time value of an option

Table 6–1   Effects of Buying a Call Option

July 1, buy 1 call option on December corn
    $2.40 Strike Price, 15 cents premium
    option is at the money

Effect of Price Increase

1) Price increase to $2.70/bushel
    EXERCISE OPTION RECEIVE long December corn
    futures position @ $2.40/bushel; sell December corn
    futures @ $2.70/bushel
    $0.30/bushel gain less 15 cents per bushel premium
    = $0.15/bushel net gain

2) Price increase to $2.90/bushel
    EXERCISE OPTION RECEIVE long December corn
    futures position @ $2.40/bushel; sell December corn
    futures @ $2.90/bushel
    $0.50/bushel gain less 15 cents per bushel premium
    = $0.35/bushel net gain

Effect of Price Decrease

1) Price decreases to $2.00/bushel
    Let option expire
    Net loss = $0.15 (the premium)

2) Price decrease to 1.80/bushel
    Let option expire
    Net loss = $0.15 (the premium)

Buying Puts and Calls

Buyers of options have at least seven different strike price/premium combinations to select from and usually many more depending upon how long the option contract has been trading. Speculative buying of options is attractive to investors because the most that can be lost due to trading is the amount of the premium. Option buyers have the right but not the obligation to exercise the option so rational traders will only exercise those options that have a gain, not a loss. *Option buying truncates the loss of trading at the premium level.* However, gains are unlimited. Buying of options is attractive to risk averse traders and those that must know in advance what the loss factor will be for any speculative investment.

Table 6–1 reveals the limited loss factor and the unlimited gains associated with buying a call on December corn. The example in Table 6–1 shows two price increases and two price decreases. Notice that no matter what price December corn drops to, the most the buyer of the call can lose is the 15-cent premium. Similarly, as price increases the buyer gets the full amount of the price increase, less the premium. Speculators who buy call options have thus limited their total loss to the amount of the premium in bear markets but will participate in all of the bull market price increase, less the premium. Figure 6–6 shows the truncation of losses and unlimited gains potential. Notice that the most the buyer can lose is the 15-cent

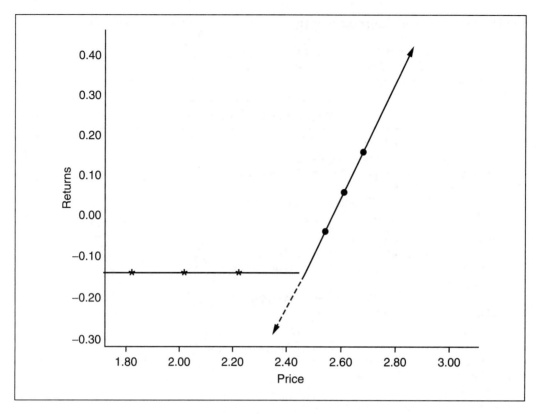

Figure 6–6   Returns from buying a call option

premium, no matter how low the price drops. Likewise, the solid sloping line up and to the right shows that once the price moves above the strike price of $2.40 per bushel, gains will occur for the buyer. The buyer will breakeven when prices move enough to cover the 15-cent premium ($2.55 per bushel). The short dashed line sloping down and to the left represents the loss potential of holding a long futures position at $2.40 per bushel. Losses are potentially unlimited with a futures position, but holding an option on having a long futures position (call) limits the loss. *Buyers of call options are speculating that the price of the underlying futures contract will increase.* If it does increase they can exercise the option and capture the price movement. If prices remain the same or decrease, they can simply let the option expire.

A speculator who thinks prices will decrease would buy a put option. Table 6–2 has the results of both a price decrease and increase with a put option. Just like with a call option, losses are truncated at the premium, but gains remain unlimited. Figure 6–7 reveals the gains associated with a price movement down from $2.40 per bushel but no matter how high prices increase, the loss is fixed at the premium level of $0.15 per bushel. The dashed line represents the potential loss if a short futures was held versus buying a put option. *Buyers of put options are speculating that the price of the underlying futures contract will decrease.*

Buying either a put or call limits the maximum loss to the amount of the premium and provides for unlimited gains. This is possible because an option buyer has the right but not

Table 6–2   Effects of Buying a Put Option

July 1, buy 1 put option on December corn
$2.40 Strike price, 15 cents premium
option is at the money

Effect of Price Increase

---

1) Price decreases to $2.70/bushel
Let option expire
Net loss = $0.15 (the premium)

2) Price decrease to $2.90/bushel
Let option expire
Net loss = $0.15 (the premium)

Effect of Price Decrease

---

1) Price increase to $2.00/bushel
EXERCISE OPTION, receive a short December corn
futures position @ $2.40/bushel; buy December corn
futures @ $2.00/bushel
$0.40/bushel gain less 15 cents per bushel premium
= $0.25/bushel net gain

2) Price increase to $2.90/bushel
EXERCISE OPTION, receive a short December corn
futures position @ $2.40/bushel; buy December corn
futures @ $1.80/bushel
$0.60/bushel gain less 15 cents per bushel premium
= $0.45/bushel net gain

---

the obligation to receive a futures contract. If a trader had directly purchased or sold a futures contract rather than an option on the futures contract, their maximum loss and gain are unlimited. Buying options allows traders to have their cake and eat it too—for the price of the premium.

Writing Puts and Calls

To **write** an option is to accept the responsibility of performing in case the buyer decides to exercise the option. Writers of options have the obligation to perform whereas buyers have the right but not the obligation to perform. The buyer pays a premium for the right and the writer gets the premium for taking on the obligation. The most income that a writer will get is the premium and the maximum loss is unlimited. *Option writing truncates the gains of trading at the premium level.* Why would any rational trader accept limited gains and unlimited losses? The concise answer is that writers have a better chance of earning limited gains versus facing an unlimited loss. Buyers of options must guess correctly which direction that price will move before they can earn any positive income. But if prices don't move at all or move in the other direction, the option will be allowed to expire. Enter the writer.

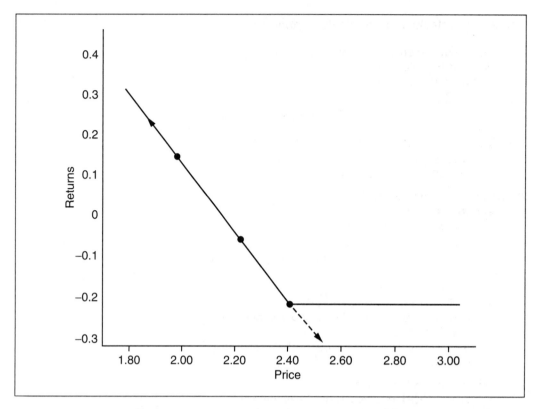

Figure 6–7   Returns from buying a put option

Writers will collect their premiums if prices don't move and/or move in the opposite direction of what buyers want. Thus, writers generally have a higher probability of earning a limited income—the old "a bird in the hand is worth more than two in the bush" idea. Consequently there is no shortage of writers of options.

Writers of options on futures must decide if they want to write the option **covered** or **naked.** A covered writer will have the underlying futures contract that the option is written against. A naked writer does not have the underlying futures and is promising to provide one in case the buyer exercises the option. Writing an option covered or naked will depend upon what the writer's attitude is toward price movements.

Writing Puts

Table 6–3 shows the effects of writing a put naked with both a price decrease and a price increase. The writer gains the maximum possible (the premium) when the price of December corn futures increases because the put option has become worthless to the buyer; thus, he allows it to expire. However, when the price decreases the buyer exercises the option and the writer has to perform. Since the option was written naked, there is no futures position to provide to the buyer. The clearing company of the exchange where the futures position was placed will create a short futures position at the strike price and give it

Table 6–3   Effects of Writing a Put Naked

Price increases to $2.80 (Buyer lets the option expire)	Seller gains the premium 5 cents/bushel
Write a Put December Corn Strike price @ $2.40/bushel (at the money) Premium 5 cents/bushel	
Price decreases to $2.00/bushel (Buyer exercises the option) Seller is assigned a long (buy) position in December corn futures at $2.40/bushel; seller offsets the buy position by selling a December corn futures @ $2.00/bushel (Loss of $0.40/bushel, gains the 5 cent/bushel premium for a net loss of $0.35/bushel)	Seller loses 35 cents/bushel

to the option buyer. One half of a futures position cannot be created, so the writer gets assigned the opposite side (long). Writers of naked puts are bullish because they lose money when the price falls and make the maximum possible when prices rise.

Table 6–4 provides the results of writing puts covered. Notice that the effects are just the opposite of writing puts naked. Covered put writers are bearish because they lose money when the price increases and earn the maximum possible when prices decrease. The covered writer must pass along the futures contract to the buyer of the option when it is exercised complete with all of the paper profits the futures position has accrued. That, of course, is the reason the buyer exercises the option—it has profits. However, the

Table 6–4   Effects of Writing a Put Covered

Price increases to $2.80/bushel (Buyer lets the option expire) Writer offsets futures position by buying December corn futures @ $4.80/bushel; loses $0.40/bushel but gains the premium of 5 cents for a net loss of $0.35/bushel	Writer loses 35 cents/bushel
Writes a put December corn Strike price @ $2.40/bushel (at the money) Premium 5 cents/bushel simultaneously sells a December corn futures @ $2.40/bushel to cover option; posts margin	
Price decreases to $2.00/bushel (Buyer exercises the option) Writer passes the sell position in December corn futures to buyer Writer gains the premium	Seller loses 35 cents/bushel

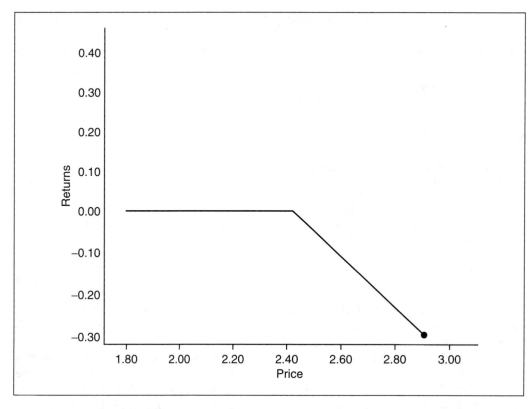

Figure 6–8   Returns from writing a put option covered

writer still earns the maximum possible—the premium. *A put option writer will cover the option if he or she is bearish and leave the option uncovered (naked) if he or she is bullish.* This concept is easily visualized by looking at Figures 6–8 and 6–9. The covered writer earns the premium regardless of how low prices go but loses money as prices increase (Figure 6–8). When the option is naked, just the opposite occurs—the premium is earned regardless of how high prices rise, but losses occur as prices dip below the strike price as illustrated in Figure 6–9.

Writing Calls

A call writer who is bullish will write the call covered, as shown in Table 6–5. The writer earns the maximum possible when prices increase and loses when prices decrease. The writer must pass along all of the paper profits in the futures position to the buyer, but earns the premium. When the market goes down, the writer must offset the losing futures position.

If a call writer is bearish, he or she will simply leave the option naked. If prices fall, as shown in Table 6–6, the buyer will let the option expire and the naked writer earns the premium. However, if prices increase, the buyer will exercise the option and receive a long futures position from the futures clearing corporation and the writer will be assigned the opposite (short) side and therefore incur a loss when the futures position is offset. *A call option*

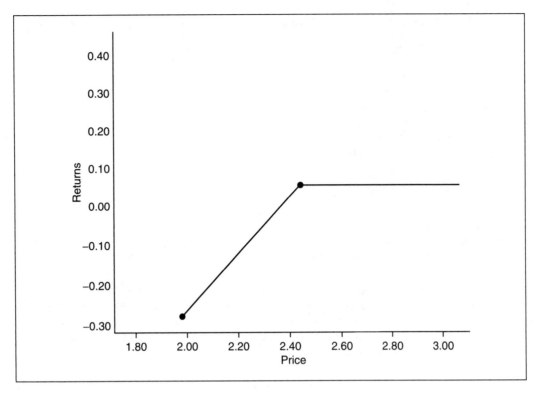

Figure 6–9   Returns from writing a put option naked

Table 6–5   Effects of Writing a Call Covered

Price increases to $2.80/bushel (Buyer exercises the option) Writer passes the buy position in December corn futures to buyer Writer gains the premium	Seller gains the premium (5 cents/bushel)
Writes a call on December corn Strike price @ $2.40/bushel (at the money) Premium 5 cents/bushel simultaneously buys a December corn futures @ $2.40/bushel to cover option; posts margin	
Price decreases to $2.00/bushel (Buyer lets the Option expire) Writer offsets the long futures position by selling a December corn futures @ $2.00/bushel for a loss of $0.40/bushel but gains the premium of 5 cents for a net loss of $0.35/bushel	Seller loses 35 cents/bushel

Table 6–6   Effects of Writing a Call Naked

Price increases to $2.80/bushel
(Buyer exercises the option)
Seller is assigned a short (sell) position in
December corn futures @ $2.40/bushel
Seller offsets the sell position by buying a
December corn futures @ $2.80/bushel
Loss of $0.40/bushel, gains the 5 cent
premium for a net loss of $0.35/bushel

Write a call on December corn
Strike price @ $2.40/bushel (at the money)
Premium 5 cents/bushel

Price decreases to $2.00/bushel
(Buyer lets the option expire)   Seller gains the premium (5 cents/bushel)

writer will cover the option if he is bullish and leave the option uncovered (naked) if he is bearish. Figure 6–11 displays the constant premium income regardless of how low prices dip and the losses that occur as prices increase when the option is naked. When the option is covered as shown in Figure 6–10, premium income is earned as prices increase and losses occur when prices decrease below the strike price.

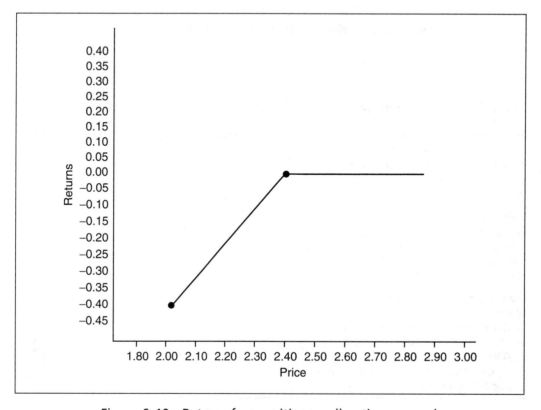

Figure 6–10   Returns from writing a call option covered

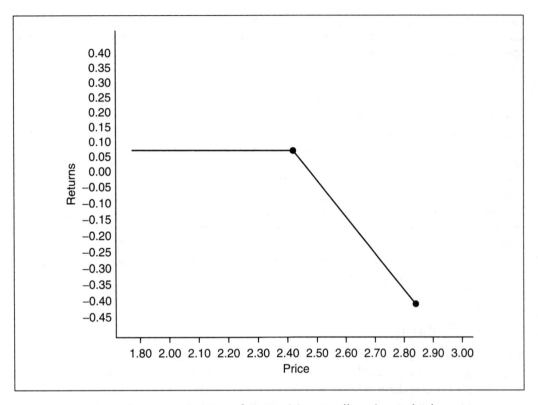

Figure 6–11   Returns from writing a call option naked

A Note on Writing versus Buying Options

If an option writer is bullish, he can write a call covered or write a put naked. Go back to Figures 6–9 and 6–10 and notice that they are identical in terms of the pattern of returns—writing a call option covered has the same return profile as writing a put option naked. *They are mechanically different, but alike financially.* Likewise, if they are bearish, they can write a call naked or write a put covered. If the writer guesses correctly, he will earn the maximum possible—the premium. If he guesses wrong, he will be subject to unlimited losses. However, since options on futures are retradable, the wise option writer will trade out of the option with only limited losses. When the term covered is used, most rational individuals will assume that some amount of risk has been managed or covered. Yet one of the paradoxes of option writing is the meaninglessness of the term covered as it pertains to risk management. The previous examples clearly show that when an option is written covered, regardless of whether or not it is a put or call, the writer can have unlimited losses if prices move in the direction opposite of what the writer expected. *Thus, writing options covered or naked is simply done to reverse the direction of price expectation and not done for any risk management reason.*

Therein lies the problem with using option writing as a price risk management tool—losses are potentially unlimited, but gains are strictly limited to the premium regardless of whether or not the option was written covered or naked. The concept of counterbalance requires that the loss in one market or financial instrument be compensated by gains in another, and vice versa. Yet that is impossible when written options are used because gains are truncated while losses are unlimited. Consequently, the use of written options as a stand-alone risk management tool is impossible. Only when written options are combined into a more complex strategy can they have some value in risk management. *Sadly, written options are widely touted as a hedging strategy when in fact they have very limited value as a hedge.*

Buying options is a valid risk management tool because losses are limited and gains are unlimited. The concept of counterbalance holds with option buying because they behave just like insurance—losses are covered, but gains are not limited. The insurance policy is used when necessary, otherwise it is not. Written options, however, assume the role of the insurance provider—losses can be unlimited and gains limited to premiums. An insurance policy holder (the option buyer) has shifted the risk of financial losses from an event (managed the risk) to the insurance company (the option writer). The insurance company has absorbed the risk, not managed the risk, which is exactly what option writing does. Insurance companies contract with other insurance companies (re-insurers) to partially manage their risk and option writers can develop sophisticated strategies to offset or pass some of their risk to others. But, the initial act of writing the option, selling the auto insurance policy, or writing the fire insurance policy is a risk acceptance strategy, not a hedge or a risk management/minimization plan.

Option Hedging

Because gains are limited with written options, only the buying of options will be used in simple hedging. Since the buying of options involves the right but not the obligation to have a futures position, hedging with options creates *the right but not the obligation to hedge with a futures contract.* With a simple futures hedge, when prices moved in favor of the cash position, the futures position lost an approximately equal amount. However, with an option hedge the holder can always let the option expire and not take a futures position. It is this simple characteristic that makes options very popular as hedges.

Put Option Hedging

A corn producer is concerned that the price of corn will decrease during the growing season and be at a very low level when the corn is harvested in the fall. A proper futures hedge would be to sell a corn futures contract during the growing season. The short futures position would gain in value if the corn price decreased. The corn producer has managed the risk of price change. The corn producer could also hedge with options. Instead of selling a futures contract, an option hedge for the corn producer would involve buying a put option on corn futures. A put is the right but not the obligation to sell a corn futures contract. Thus, the corn producer has bought the right but not the obligation to hedge. If corn prices go down, the producer will *exercise the option and have a futures hedge.* On the other hand, if prices go up and the hedge is not needed, the producer can let the option expire. Table 6–7 shows the effects of both a price decrease and a price increase.

The corn producer in the example would never receive anything less than $2.35 per bushel regardless of how low prices went. Yet when the cash price increased, the option can be allowed to expire and the producer will only lose the premium. This put hedge sets a floor below which the net effect to the producer will be fixed at the floor, but the upside is unlimited and reduced only by the amount of the premium. The producer simply takes the strike price and subtracts the premium amount to set the floor. Since numerous combinations

Table 6–7 Effects of a Put Option Hedge

Cash	Option
July 1, corn crop growing current cash corn price @ $2.50/bushel	Buys a put option on December corn futures at a strike price of $2.60/bushel and premium of $0.15/bushel (at-the-money option)

Price Decrease

Cash	Option
Oct 1, harvests and sells corn locally @ $2.00/bushel	Exercises the option; receives a sell position in December corn futures @ $2.60/bushel; buys December corn futures @ $2.10/bushel; gains $0.50/bushel
	Less $\dfrac{\$0.15 \text{ premium}}{\$0.35 \text{ net gain}}$

Net hedged option price = $2.00 + $0.35 = $2.35/bushel.

Price Increase

Cash	Option
Oct 1, harvests and sells corn locally @ $3.00/bushel	Lets option expire Loss of premium = $0.15/bushel

Net hedged option price = $3.00 – $0.15 = $2.85/bushel

of strike prices and premiums are available at any one time, the producer can adjust the floor to the level that he wants.

Call Option Hedging

A large feed company has forward sold pre-mixed feed for delivery in two weeks. The major ingredient is corn valued at $2.50 per bushel even though the company will not buy the corn until a day before delivery. The company has the risk that between the forward pricing of the corn at $2.50 per bushel and the actual purchase of the corn, that the price will increase and thus squeeze and/or eliminate the profit margin. Table 6–8 shows the effects of using a call option to hedge the risk.

The company's ceiling price was $2.65 per bushel. The company would never pay any more than that amount for corn regardless of how high prices increase. However, when prices fell, the company actually paid less than $2.50 per bushel for the corn. Call option hedging sets a ceiling on prices. In this particular case, if the company has any market power, they could have valued the corn in the ration at the ceiling price of $2.65 per bushel rather than $2.50 per bushel. The company simply would take the current price of $2.50 per bushel and add the call option premium of 15 cents per bushel and value the corn at $2.65 per bushel. The cost of the option is thus passed on to the buyer of the feed. Depending upon

Table 6–8   Effects of a Call Option Hedge

Cash	Option
Jan 2, forward sells pre-mixed feed for delivery in 2 weeks with corn valued @ $2.50/bushel	Buy a call option on March corn futures at a strike price of $2.60/bushel Premium of $0.15/bushel (at-the-money option)
Price Decrease	
Jan 15, buys corn @ $2.30/bushel	Lets the option expire Loss of premium of $0.15/bushel

Net hedged option price = $2.30 + $0.15 = $2.45/bushel

Price Increase	
Jan 15, buys corn @ $2.90/bushel	Exercises the option; receives a buy position in March corn futures @ $2.60/bushel; sell March corn futures @ $3.00/bushel
	Gains $0.40/bushel
	Less $\dfrac{\$0.15/\text{bushel premium}}{\$0.25/\text{bushel gain}}$

Net hedged option price = $2.90 – $0.25 = $2.65/bushel

how price sensitive the buyer is will determine whether or not the seller could include the price of the option in the valuing of the feed.

Floors and Ceiling with Option Hedging

Using the two previous examples in Tables 6–7 and 6–8, Figures 6–12 and 6–13 illustrate the concept of price floors and ceiling with option hedging. Figure 6–12 shows that with a put option hedge, the lowest net hedged price (price floor) the corn producer will receive is $2.35 per bushel regardless of how low market prices get. Likewise in Figure 6–13 the highest price the feed company will pay for corn (price ceiling) is $2.65 per bushel no matter how high prices get. The figures show the real attraction of hedging with options—price protection, but also the ability to have a gain when prices move favorably. With Figure 6–12 the corn producer clearly got a price floor protection but was also able to keep all of the market price increase less the cost of option premium. The same goes for the feed company with call options as revealed in Figure 6–13. The feed company received the protection of a price ceiling but got the advantage of lower market prices when market prices decreased.

Hedging with futures certainly provides a price floor for a corn producer, but at the same time it furnishes a price ceiling. In other words, the producer is protected against

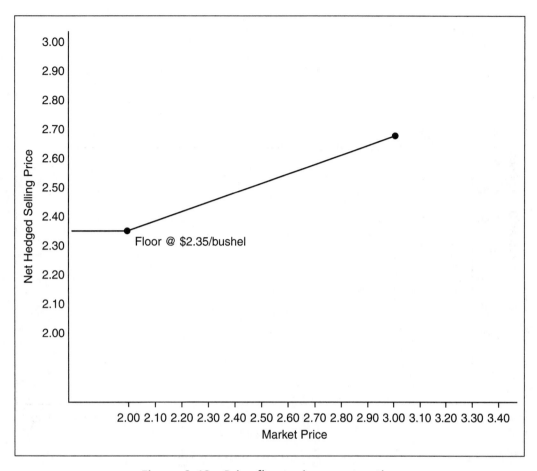

Figure 6–12   Price floor using a put option

price movements both favorable and unfavorable. Option hedging provides either a floor or ceiling for unfavorable price movements, and lets the hedger take advantage of favorable movements. Of course this attribute of options has a cost—the premium.

Hedging to a Certain Price/Cost

Similar to hedging price floors and ceilings is the use of options as a hedging tool to achieve a certain price or cost coverage. Since options have several strike prices to pick from when structuring a hedge, it is important to establish criteria for determining which strike price to hedge. Put option hedgers can subtract the premium from the strike price and derive the price floor, or also commonly called the **target price.** Table 6–9 shows hypothetical strike prices, premiums, and target prices. The formula for a target price is

$$\text{target price} = \text{put strike price} - \text{premium}$$

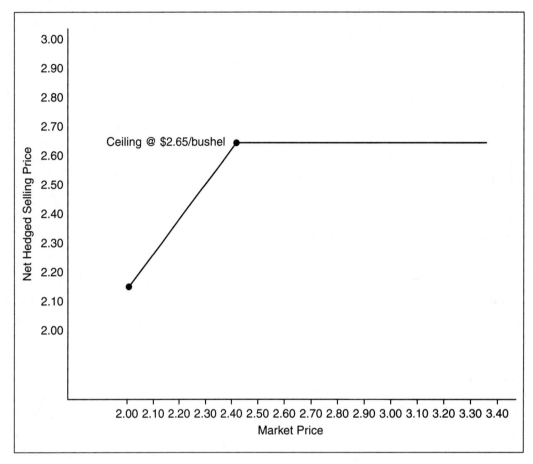

Figure 6–13 Price ceiling using a call option

Table 6–9 Target Prices for Feeder Cattle Put Options

Put Strike Price /cwt.	Premium /cwt.	Target Price /cwt.
$100.00	$3.10	$96.90
$99.00	$2.15	$96.85
$98.00	$1.35	$96.65
$97.00	$0.50	$96.50
$96.00	$0.45	$95.55
$95.00	$0.25	$94.75
$94.00	$0.15	$93.85

Table 6–10   Target Prices for Feeder Cattle Call Options

Put Strike Price /cwt.	Premium /cwt.	Target Price /cwt.
$100.00	$0.15	$100.15
$99.00	$0.25	$99.25
$98.00	$0.45	$98.45
$97.00	$0.50	$97.50
$96.00	$1.35	$97.35
$95.00	$2.15	$97.15
$94.00	$3.10	$97.10

Call option hedgers are concerned about a price ceiling and thus if the premium is added to the strike price a target cost can be calculated as shown in Table 6–10. The formula for target costs is:

target cost = call strike price – premium

Both Table 6–9 and 6–10 use the same product, feeder cattle, and the same set of strike prices and premiums (reversed for puts and calls). Feeder cattle can be a final product for a rancher and thus need downside price protection and they can also be an input for cattle feeders and thus need upside price protection. Feeder cattle options thus provide a clear concept of both target prices and costs.

Target Price Hedging

Consider a rancher who is trying to decide whether to sell his calf crop as stockers or to carry them to heavier weights and sell them later as feeder cattle. The rancher can now use the information in Table 6–9 as a management tool in setting his criteria for making the decision to carry the animals to heavier weights. The rancher calculates that if he transforms the stocker cattle into feeder cattle by carrying them longer on grass and supplemental feed, his estimated cost of production is $95 per hundredweight. By using the information in Table 6–9, he has very valuable information *in advance of making the decision*. All of the information about the target prices in Table 6–9 is available in advance. The rancher, for example, could hedge by buying a one-step out-of-the-money put option at a strike price of $96 for a premium of 45 cents per hundredweight for a calculated target price of $95.55 per hundredweight as shown in Table 6–11. If the rancher's estimated cost of production is correct and he hedges with the $96 put option, his gross profit will be 55 cents per hundredweight. If his estimated final feeder weight is 700 pounds, his gross profit per head will be $3.85. This value is, of course, a minimum. If feeder cattle prices increase during the time the rancher held the animals, the option hedge could be allowed to expire and the rancher would gain from the cash price increase as revealed in Table 6–11. The target price hedge becomes the price floor. The rancher could also hedge with a $100 strike price, in which he would have a target price of $96.90 or a gross profit of $1.90 per hundredweight or $13.30 per head. The $100 strike price hedge involves a larger insurance premium than the $96 strike price and thus a larger negative cash flow, but it guarantees a larger profit potential—a classic management decision—higher guaranteed profits, but at a cost.

Table 6–11 Put Target Price Hedge

Cash	Financial
Decides to carry stockers to heavier weights (feeder cattle cash price $97/cwt.)	Buys put @ $96/cwt. Pays premium of $0.45/cwt.
	Target Price = $96 – $0.45 = $95.55/cwt.

Cash Price Decrease

Net hedged price = $80 + $15.45 = $95.55

Cash Price Increase

Sell feeder cattle for $105/cwt.	Let option expire Lose premium of $0.45/cwt.

Net hedged price = $105 – $0.45 = $104.55/cwt.

Target Cost Hedging

A cattle feeder needs feeder cattle as an input. Table 6–10 shows the array of possible costs of feeder cattle relative to the initial strike price used as a hedge. A cattle feeder, for example, might have the potential to forward sell the finished fed cattle for a certain price and then would estimate the feed and other costs involved, set a target profit margin, and determine that they could pay no more than $98 per hundredweight for the feeder cattle. Should the cattle feeder proceed? The target cost array shows several possible call option hedges that would satisfy the feeders' goals. If the cattle feeder hedges by buying the $97 strike price call, the target cost of feeders will be $97.50 per hundredweight as shown in Table 6–12. The cattle feeder achieves the profit margin she wanted with a feeder cattle cost of $98 per hundredweight and the hedge with the call option beat the price by fifty cents per hundredweight. Assuming a 700-pound feeder, the cattle feeder will add an additional $3.50 per head to the estimated profit margin. The hedge also allowed for the cattle feeder to capture a lower feeder cattle cost when prices move lower by letting the option expire as Table 6–12 reveals. The cattle feeder could also hedge with a $94 strike price at a substantial premium of $3.10 per hundredweight for a target cost of feeder cattle at $97.10 per hundredweight. The cattle feeder would gain a reduced cost of 40 cents per hundredweight for an additional cost of $2.60 per hundredweight. Given that the target price of feeders is achieved with the higher strike price and lower premium, the cattle feeder might not opt for the lower target cost with its corresponding higher negative cash flow. *However, that is a management decision and can be made prior to the production decision using the concept of target cost hedging.*

Calculating target prices and costs can be done in advance of many business, financial, and production decisions and thus can be very powerful as a management tool. Consider the simple process of placing cattle on feed. If a cattle feeder calculates that they can pay no more than $85 per hundredweight for the feeder cattle and assuming the prices in Table 6–10 are the only hedging choices, a cattle feeder has a potent bit of information. No possible hedge exists with call options that would guarantee a profit. The cattle feeder would have to lower their profit expectations and or find cost savings elsewhere, or simply not feed cattle.

Using the target price and costs arrays also leads to another potential type of option hedging—partial coverage (also known as worst-case or catastrophe hedging).

Table 6–12   Call Target Cost Hedge

Cash	Financial
Current feeder cattle price @ $98/cwt.	Buys a call @ $97/cwt. Pays premium of $0.50
	Target cost = $97 + $0.50 = $97.50

Cash Price Increase	
Buys feeder cattle @ $101/cwt.	Exercise option
	Receives long feeder cattle futures @ $97/cwt.
	Sells @ $101/cwt.
	Gross margin of $4/cwt. less premium of $0.50/cwt.
	Net margin = $3.50/cwt.
Net hedged price = $101 − $3.50 = $97.50/cwt.	

Cash Price Decrease	
Buys feeder cattle @ $93/cwt.	Lets option expire
	Lose premium of $0.50/cwt.
Net hedged price = $93 − (−$0.50) = $93.50/cwt.	

Partial coverage hedging is the idea that only a certain level of protection is desired. This could be because of the cost of full coverage or the hedger may want to speculate that prices will move in their favor and thus they want only a worst-case type of coverage.

Partial Coverage Hedging

In the example in Table 6–11, a rancher is looking at carrying his stockers to feeder cattle weights and hedges with a put option and achieves a target price of $95.55 per hundredweight. In that example the rancher calculated that he needed to receive at least $95 per hundredweight. Assume instead that the rancher's total cost of production was $97 per hundredweight and that the rancher's fixed costs were estimated to be $93 per hundredweight. The rancher is generally bullish on feeder cattle prices, but wants to make sure he can make his land mortgage payments. He wants protection for only the fixed cash portion of the costs. Using the information in Table 6–9, the rancher could hedge with the $94 three-step out-of-the-money put option with a target price of $93.85 per hundredweight. This hedge would allow the rancher to cover his fixed costs and have 85 cents per hundredweight extra as a contribution to variable costs as a minimum. The rancher has partial coverage of his total costs with the possibility of more if his bullish expectations pan out, as shown in Table 6–13. In the example in Table 6–13, the rancher is protected at a price of $93.85 per hundredweight that doesn't cover all of his costs, but does cover his fixed costs, but in the event prices increase, he can take advantage of that as well.

The astute reader can clearly see how partial coverage hedging with options relates to car or home insurance. Full replacement insurance costs more than partial coverage, and likewise hedgers with options can use the concept to provide financial protection at various levels.

Table 6–13 Partial Coverage Hedge

Cash	Financial
Decides to carry stockers to heavier weights (feeder cattle cash price $97/cwt.)	Buys put @ $94/cwt.
	Pays premium of $0.15/cwt.
Needs $97/cwt. to cover all costs and $93/cwt. to cover fixed costs.	Target price = $94 – $0.15 = $93.85/cwt.

Cash Price Decrease	
Sells feeder cattle for $80/cwt.	Exercise option
	Receives short feeder cattle futures @ $94/cwt.
	Buys back @ $80/cwt.
	Gross margin of $14/cwt. less premium of $0.15/cwt.
	Net margin = $13.85/cwt.
Net hedged price = $80 + $13.85 = $93.85/cwt.	

Cash Price Increase	
Sells feeder cattle for $105/cwt.	Lets option expire
	Lose premium of $0.15/cwt.
Net hedged price = $105 – $0.15 = $104.85/cwt.	

How much property and casualty insurance coverage a business needs to carry is related to many factors—how much debt the business carries, owners' attitude toward risk, forecasts of risky events occurring, cash flow, and premium costs. Likewise, how much price insurance a business should carry is related to many of the same factors. Option hedging provides an almost limitless array of possible partial coverage opportunities for agribusinesses.

Option Multiple Hedging

To properly use option as a hedging tool requires knowing that options are double derivatives. Options on futures derive their value from the underlying futures contract that in turn derives its value from the underlying cash market. The relationship between the cash market price and the futures market price, as explained in Chapter 4, is called basis. Knowledge of basis and basis movements is critical to effective price risk management using futures contracts. The same is true for options. The relationship between the value of the option (the premium) and the underlying futures price is called *delta* (expressed as Δ). The formula for delta is

delta = change in the option premium/change in futures price

Deltas range from zero to one. A delta of zero means that the option premium did not change when there was a change in the price of the underlying futures contract. Deep out-of-the-money options will have deltas close to or at zero. A delta of one implies that for every dollar

Table 6–14   Impact of Delta on Effectiveness on Option Hedging

Cash	Futures	Options
Buys @ $3.00	Sells @ $3.00	Buys put strike price @ $3.00
		Pays premium $0.05, Delta 0.5
Price decrease		Futures price change = $0.10
		Option premium <u>increases</u>
		($0.10 × 0.5 (delta)) = $0.05
<u>Sells @ $2.90</u> −$0.10	<u>Buys @ $2.90</u> +$0.10	<u>Sells Put for 10¢ (5¢ + 5¢)</u> +0.05
Price increase		Futures price change = $0.10
		Option premium <u>decreases</u>
		($0.10 × 0.5 (delta)) = $0.05
<u>Sells @ $3.10</u> +$0.10	<u>Buys @ $3.10</u> −$0.10	<u>Put premium = 0</u> −$0.05

change in the underlying futures price, the option premium changed a dollar as well. Deep in-the-money options have deltas that approach one or are one. As an option moves from at the money to in the money, intrinsic value is picked up and deltas start increasing and vice versa.

Deltas impact the concept of counterbalance. The idea of proper hedging stems from having the potential loss in the cash market value offset with a similar gain with a financial instrument used as a hedge. If the hedge involves an option with a delta of 0.5, then for every dollar the futures price changed, the option premium only changed by 50 cents, thus a major mismatch in counterbalance. Table 6–14 shows the impact of the change in the value of the option with a price decrease and increase. In the example, the futures hedge fully protected the cash position with either a price increase or decrease (assuming a zero basis and zero basis change). The option hedge only protected half of the cash price movement when prices moved down because the option premium changed by 50 percent.

To provide for dollar equivalency the option hedge in the example in Table 6–14 would need to be twice as large as the futures hedge. The process of achieving dollar equivalency between futures hedges and option hedges is called the *multiple*. The formula is:

$$\text{multiple} = 1/\text{delta}$$

With a delta of 0.5 in Table 6–14, the multiple would be 2 (1/0.5). The multiple is simply the inverse of delta. In the example in Table 6–14, two puts would need to be purchased instead of just one.

A proper multiple option hedge will mimic the effects of a futures hedge when the price moves against the cash position as it did when it decreased in the example in Table 6–14. In the example in Table 6–14, a multiple option hedge would be two put options purchased, for a total gain on the options of hedge of 10 cents (5 cents × 2). However, as the example in Table 6–14 also shows, when prices move in favor of the cash position, the option hedge will likewise mimic the futures hedge and the loss of 5 cents would double to 10 cents.

The central idea behind deltas and multiples is to understand how the value of the option as a hedge changes as the futures price changes. Hedgers that desire to have dollar

Table 6–15 Scaling an Option Hedge Down

Cash	Financial
Buy corn (5,000 bushels) @ $3/barrel	Buy put options (10) @ $2.80 strike price (out of the money)
	Premium = $0.02, delta = 0.1
	Multiple = 10
Corn price goes to $2.80/bushel	Delta = 0.5
	Multiple = 2
	Scale down
	Hedge
	Sell 8 put options @ $2.80 strike price (now at the money)
	Premium $0.05
	$0.05 − $0.02 = $0.03 gain
	$0.03 × 40,000 (8 × 5,000) = $1,200
Sell corn @ $2.75/bushel	Sell 2 puts @ $2.80 strike price (now at the money)
−$0.25/bushel × 5,000 = −$1,250	Premium of 6¢
	$0.06 − $0.02 = $0.04 gain
	$0.04 × 10,000 (2 × 5,000) = $400
	$1,200 + $400 = $1,600 Total Gain
Net hedged price = $1,600 − $1,250 = $350/5,000 = $0.07/bushel	

equivalency throughout the hedging period will need to *scale the hedge.* As the futures price changes during the hedge period, so will the delta, and thus the multiple. If the hedge is not scaled (adjusted for size) then the hedge will not achieve dollar equivalency. The example in Table 6–15 displays a hedge that started out with deep out-of-the-money options with a delta of 0.1, but as the futures prices moved, and moved the options to at the money, the delta rose to 0.5. The hedge needs to be scaled down by offsetting. The end result of the scaling process in Table 6–15 produced a net hedged price of $2.82 per bushel which is close to the $2.80 per bushel strike price that was initially purchased as the level of desired protection.

Table 6–16 shows an option hedge that was scaled up to reflect a decreasing delta as the market price moved up and thus against a put position. The level of protection was initially set at $2.90 with the beginning strike price and the final net hedged price of $3.07 per bushel is considerably better since the cash price increased.

The process of scaling option hedges attempts to achieve *delta neutral hedges.* A true delta neutral hedge would achieve perfect dollar equivalency and thus a perfect hedge. A perfect hedge will mitigate all price risk, but will not allow any profit potential. Most option hedgers, therefore, are aware of the importance of delta as a measure of how option premiums change as the underlying futures price changes, but generally shun trying to achieve delta neutral hedges.

Table 6–16 Scaling an Option Hedge Up

Cash	Financial
Buy corn @ $3/bushel	Buy put options (4) @ $2.90 strike price (out of the money)
	Premium = $0.04, delta = 0.25
	Multiple = 4
Corn price goes to $3.20/bushel	Buy put options (6)
	Premium = $0.02, delta = 0.1
	Multiple = 10
Sell corn @ $3.25/bushel	Sell 10 put options
$-$0.25/bushel $\times$ 5,000 = $-$1,250	Premium = 0.01
	Buy @ 4¢ – 1¢ = – 3¢ $\times$ 20,000 = –$600
	Buy @ 2¢ – 1¢ = – 1¢ $\times$ 30,000 = $\underline{-\$300}$
	–$900

Net hedging gain = $1,250 – $900 = $350/5,000 = $0.07/bushel.

Net hedged price = $3 + $0.07 = $3.07/bushel

A Final Word on Deltas

Deltas of necessity have to be calculated after the fact, thus they are always historic and static. Markets are dynamic and therefore deltas will constantly be changing. Option hedges, consequently, are in constant need of being scaled. Scaling increases the cost of hedging by having to pay additional brokerage fees. Furthermore, as options hedges are scaled they more closely pattern a futures hedge. If the end result of an options hedge is similar to a futures hedge, then a futures hedge would be superior since option scaling involves higher brokerage fees. A wise options hedger is aware of the importance of deltas as well as their limitations.

Synthetic Futures Hedging

Option writing when combined with option buying can, if constructed properly, mimic a futures hedge. If a short futures hedge is desired, the combination of buying a put and simultaneously selling a call on the same futures contract will exactly reproduce the effects of a futures hedge as shown in Table 6–17. The opposite is true for a long futures hedge—buy a call and sell a put. This combination will exactly mimic the long futures hedge. In the example in Table 6–17 the futures hedge netted a price of $2.50 per bushel regardless of whether or not the price increased or decreased. Likewise, the synthetic set of options produced the exact same results.

A **synthetic futures hedge** is favored by some brokers because the client will have to pay two commissions for the options versus just one for a futures hedge. A synthetic futures hedge that has the options positions at the same strike price and roughly the same premium should be avoided. *This type of hedge simply parrots a futures hedge and costs more to place. They should be avoided.*

Table 6–17   Synthetic Futures Hedge

Cash	Futures	Options	
Buy corn @ $2.50	Sell December corn @ $2.60	Buy a put Strike price $2.60 Premium $0.10 (at the money)	Sell a call Strike price $2.60 Premium $0.10 (at the money)
	Price Increase		
Sell @ $3.00 + $0.50	Buy @ $3.10 – $0.50	Let put expire Lose $0.10	Buyer of call exercises option; receives sell position in December corn futures @ $2.50. Buy back @
			$3.10 –$0.50
		–$0.10 Net loss = $0.50	Receive $0.10 Premium – $0.40
	Net hedged futures price = $3.00 – $0.50 = $2.50	Net hedged option price = $3.00 – $0.50 = $2.50	
	Price Decrease		
Sell @ $2.00 – $0.50	Buy @ $2.10 + $0.50	Exercise put, receive sell position December corn @ @ $2.60. Buy back @	Buyer lets expire Earn premium $0.10
		$2.10 +$0.50 less Premium ($0.10) $0.40 Net gain on put plus $0.10 on call = $0.50	
	Net hedged futures price = $2.00 + $0.50 = $2.50	Net hedged option price = $2.00 + $0.50 = $2.50	

Obviously, limitless combinations of strike prices and premiums exist for the synthetic hedge. One of the most popular synthetic futures hedge involves buying an out-of-the-money put (call) and selling an in-the-money call (put) such that the hedge initially earns a net cash flow for the hedger. Table 6–18 illustrates this concept. This strategy is recommended by some brokers as a way for the hedger to earn an initial positive cash flow and thus it appears

Table 6–18   Synthetic Futures Hedge with Different Premiums

Cash		
Buy corn @ $2.50/bushel	Buy a put Strike price $2.30 Premium $0.03	Sell a call Strike price $2.30 Premium $0.23
	Net Premium Gain of $0.20 (0.23 – $0.03)	

Price Decrease

Sell @ $2.00 / bushel – $0.50	Exercise the put, receive sell futures position @ $2.30 Buys back @ $2.00 +$0.30 Less $0.03 Premium +$0.27	Buyer lets expire Gains $0.23 + $0.23

Net hedged price = $2.00 + $0.50 = $2.50   Net gain of $0.50 ($0.27 + $0.23)

Price Increase

Sell @ $3.00 / bushel +$0.50	Let put option expire Lose $0.03	Buyer exercises option Seller assigned a sell futures position @ $2.30 Buys back @ $3.00 –$0.70 +$0.23 –$10.47

Net hedged price = $3.00 – $0.50 = $2.50   Net loss $0.50

No Price Change

Sell $2.50 $0.00	Let put option expire Lose – $0.03	Buyer exercises option Seller assigned sell futures position @ $2.30 Buys back @ $2.50 –$0.20 +$0.23 +$0.03
	Net = 0 ($0.03 – $0.03)	

Net hedged price = $2.50 + $0.00 = $2.50/bushel

Table 6–19 Comparing Futures Hedges with Options Hedges

Cash	Futures	Options
Buy @ $2.50	Sell @ $2.60	Buy a put Strike price @ $2.60 Premium $0.05 (at the money)
Price Decrease		Exercise option; receive Sell position @ $2.60

Sell @ $2.00
  −$0.50
	Buy @ $2.10
	  +$0.50
		Buy back @
		  $2.40
		  +$0.50

		Loss $0.05 premium +$0.45

Net hedged futures price = $2.00 + $0.50 = $2.50
Net hedged options price = $2.00 + $0.45 = $2.45

Price Increase		Let option expire Lose premium = −$0.05

Sell @ $3.00
  +$0.50
	Buy @ $3.10
	  −$0.50	

Net hedged futures price = $3.00 − $0.50 = $2.50
Net hedged options price = $3.00 − $0.50 = $2.95

that the hedge actually makes a profit right off the bat. However, as the example shows, the effects are the same if prices go up, down, or do nothing.

To be sure, some times playing the game of putting on a synthetic hedge with differing premium values might earn the hedger a small net gain, but other combinations will cause a small net loss. As a rule of thumb, a hedger should merely hedge with a futures instead of a synthetic hedge with options. Unfortunately, the synthetic hedge is alive and well in the brokerage community and a prudent hedger should find another broker if the broker continues to recommend synthetic hedges.

Options Hedges Versus Futures Hedges

It might appear from the previous discussions and examples that options are clearly superior forms of hedging versus futures contracts; nevertheless, that is not the case. Options are a super way to hedge sometimes, and an inferior way to hedge at other times. Academics call options "**second best**" inasmuch as an option hedge is less effective than a futures hedge when prices move against the cash position. If prices move in favor of the cash position, an option hedge is more costly than not being hedged at all. The option hedge will always cost the premium and accordingly will always be "second best."

This idea of second best sounds correct and in fact is correct with one glaring fault—it can only be judged after the fact. Consider the example in Table 6–19. In the situation where the price moved against the cash position (down), the futures hedge was superior to the options hedge by the amount of the premium. When prices moved in favor of the cash position, the options hedge was superior to the futures position, but inferior to not being hedged at all—by the amount of the premium. Ergo the idea that options are always second best. But this is clearly unfair to both the hedger and options. No one knows in advance what prices will do. To say that an option hedge is second best to not being hedged at all is true, but in reality a worthless comparison. Options are used as a risk management tool and being unhedged is absorbing risk, not managing risk.

Instead of viewing options as second best, they should be considered a tool to be used in conjunction with futures contracts. *The general rules for using one tool over the other are as follows: If prices are forecasted to move against the cash position, hedge with futures contracts. If prices are forecasted to move in favor of the cash position, hedge with options.*

CHAPTER 6—QUESTIONS

1. A March corn call option has a strike price of $2.50 per bushel. For the call to be in the money, what would the underlying March corn futures have to be?

2. What is the relationship between the strike price and the premium?

3. Explain why option writing has very limited risk management potential?

4. Why is option hedging similar to price insurance but futures hedging is not?

5. What determines the value of an option?

Swaps

The objective of this chapter is to provide an introduction to swap contracts. Agriculture, unlike other industries, has not embraced swaps, yet they are an extremely powerful tool to not only manage the risk of price change, but to actually change price levels. The combination of these two persuasive tools makes the swap market a very compelling force that needs wider use and better understanding.

OVERVIEW

The swap market is a very young market. Its origin can be traced to the late 1970s as a by-product of back-to-back loans. A back-to-back loan involves two companies, each in different countries, borrowing money from local banks and then loaning money to each other. The two companies are, in essence, swapping loans. This action avoided any taxes on external capital flows (as was the case in the United Kingdom in the late 1970s) and provided access to foreign exchange. Very quickly these back-to-back loans matured into what is known today as swaps. Accordingly, the swap market is less than 25 years old. Forward contacts and options markets can be traced to ancient markets that are over 6,000 years old and modern day futures contracts are at least 150 years old. By market standards, swaps are mere children. But they are noisy and large children; recently, swap market volume exceeded 10 trillion dollars.

Swaps are used as risk management tools by banks, financial institutions, international companies, and manufacturers. Three major types of swaps constitute most of the activity: commodity, interest rates, and currency. Regardless of the type of swap, they all share similar characteristics. This chapter will focus on commodity and interest rate swaps because they are the type of swaps that agribusinesses will typically be exposed to in the marketplace. Currency swaps are very popular in international finance but pertain to very large operations that use foreign exchange transactions.

Swap Basics

Swaps are really just the exchange between two or more parties of certain types of cash flows. Rarely do swaps involve the exchange of the principal or core value, and if they do they are called equity swaps. Otherwise, swaps merely comprise the exchange of cash flows.

As with all concepts, a few basic terms must be defined. The principal or core value is called the **notional.** The notional's value is used to calculate the **service payments**

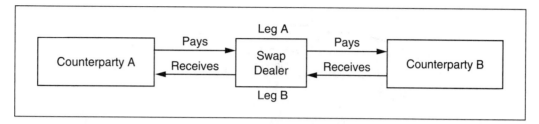

Figure 7–1   Plain vanilla swaps

that will be exchanged between the two contracting parties. The contracting parties are called **counterparties.** A swap broker will arrange the transactions between the two counterparties and receive a fee. A swap dealer will construct a deal with a counterparty and then play the role of the other counterparty until another counterparty can be booked. Swap dealers receive for their services the **bid-ask spread,** also known as the **pay-receive spread.** The bid-ask spread is simply the difference between the cash flow one counterparty pays to the dealer and what the dealer pays out to the other counterparty. The size of the spread is primarily determined by competition and the negotiation skills of the dealer. In the early 1980s it was not uncommon for swap dealers to earn a full 100 or more basis point spread on interest rate swaps, but as more dealers have entered the market, the spreads have fallen to approximately 40 basis points. Swap dealers have a professional trade association called the International Swaps and Derivatives Association (ISDA).

A basic swap is called a **plain vanilla swap** and is shown in Figure 7–1. With a plain vanilla swap, counterparty A agrees to exchange a certain amount of cash flow with counterparty B. The swap dealer that arranged the deal between the two counterparties will receive a cash flow from counterparty A and extract a certain value from it and then pass on the remainder to counterparty B (leg A in Figure 7–1). Often the dealer will do the same for the cash flow that passes from counterparty B to A (leg B in Figure 7–1). Thus the spread that the dealer receives may be from both legs of the cash flow exchange or from just one leg. The starting date for a swap is called the **effective date** and the ending date is the **termination date.** The length of time between the effective date and the termination date is called the swap **maturity** or tenor.

Commodity Swaps

The energy and metals markets use **commodity swaps** as a major tool in price risk management. The authors are unaware of any agricultural commodity that use swaps in any way. To be sure, the potential for swaps in agriculture is phenomenal, but a market has not yet been developed.

A plain vanilla commodity swap is based on a **reference price** (also known as a "fixing"). The two counterparties buy and sell the actual cash commodity in the normal market channel; however, the cash flow they agree to swap is based on a reference price so that they are both swapping values based on a common price. Reference prices are usually based on a well known location, such as Henry Hub for natural gas, or the West Texas sweet crude oil price for crude oil. Reference prices can also be a composite price such as the Cattle Fax Feeder Cattle index that is used to cash settle feeder cattle futures contracts.

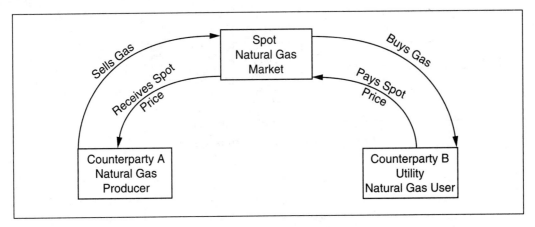

Figure 7–2   Natural gas market flows

Natural Gas Commodity Swap Example

Counterparty A is a gas producer in southeastern New Mexico. It produces natural gas and sells it into the spot market. Counterparty B is a large utility that buys natural gas in the spot market. Figure 7–2 displays the actions of both. Counterparty's A price risk is that the price of natural gas will decrease and it will earn less while counterparty B's risk is that the price of natural gas will increase and cost more to procure the product. One way for counterparty A to remove the risk of decreasing its prices is to swap the variable market price for a fixed price. Likewise, counterparty B could swap the variable market rate for a known fixed price. Both parties have a price risk that is opposite and thus they have an incentive to manage the price risk. A swap dealer could enter into a commodity swap with counterparty A as follows: counterparty A will pay to the swap dealer the Henry Hub natural gas price each Friday for 100,000 cubic feet of natural gas and in return the dealer will pay to counterparty A each Friday a fixed price of $2.50 per cubic foot based on a volume of 100,000 cubic feet. Counterparty A has swapped an unknown variable price each week for a known fixed price. Counterparty A has mitigated the risk of a price decrease. The swap dealer would offer to counterparty B the following: counterparty B will receive from the swap dealer each Friday the Henry Hub price on a volume of 100,000 cubic feet of natural gas. In return, counterparty B will pay to the dealer a fixed price of $2.55 per cubic foot on a volume of 100,000 cubic feet. Counterparty B has offset the risk of natural gas prices going up by swapping the variable rate for the payment of a fixed rate.

Figure 7–3 shows the cash flows. Notice that counterparty A received from the market a variable rate for the sale of his gas which he passes on to the dealer. The dealer pays counterparty A the fixed rate. The swap for counterparty A has converted a variable price issue to a fixed price.

Counterparty B no longer has the risk of paying a higher rate for natural gas because the swap dealer passes a variable rate price to the utility that they can use to pay for the market gas and in return they pay a fixed gas price to the dealer, in effect mitigating the risk of price change. Both counterparties still deal with the normal market channel to buy and sell gas, but the swap deal has now removed the risk of price change for both parties and the dealer gets a bid-ask spread of 5 cents per cubic foot on a weekly volume of 100,000 cubic feet or $5,000 per week.

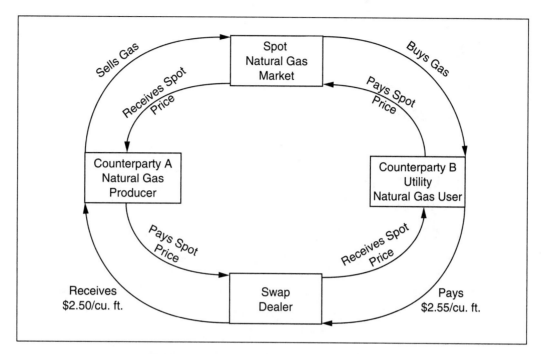

Figure 7–3   Natural gas commodity swap

Energy commodity swaps are very popular and are a very appropriate risk management tool for agribusiness firms that use large volumes of energy as a way to control costs. One of the important considerations with commodity swaps is the timing of the cash flows that will be exchanged. Energy is a natural commodity for swap contracts because the production flows of carbon-based energy is more or less continuous and likewise for end users. Consequently, producers have a continuous risk of price declines while utilities or other end users are constantly worried about price increases, and the two cash flows provide an instinctive demand to be swapped. Other commodities that are more discrete in either production or consumption need special arrangements via the swap dealer.

Agricultural Swaps

Agriculture has not embraced swaps and consequently there is no bona fide swap market in the industry. Yet agriculture has many potential swap opportunities. Elevators that gather, store, and merchandise grains and oilseeds on a regular basis have swapping opportunities with continuous grain users such as feed mills, feedlots, animal confinement feeding operations, and food processing companies. Discrete production (such as yearly crop harvests) poses a unique swapping problem, but can be handled with a **tier or ladder maturity** strategy by the swap dealer. Two agricultural potential swap examples will be discussed: (1) corn marketing operations and (2) corn production operations.

Corn Marketing Swap

Elevators buy grain from producers and other sources. They blend, dry, store, and then sell the grain to users such as feed lots, feed mills exporters, and food processors. They are constantly

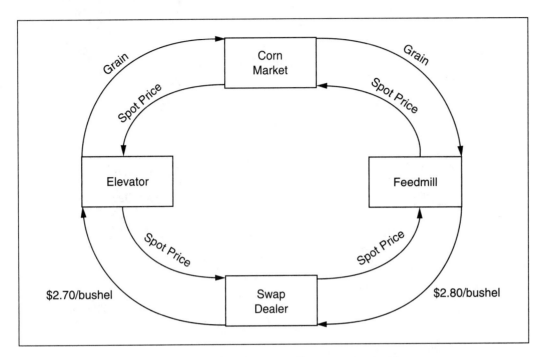

Figure 7–4   Corn commodity swap example

selling the grain into the market and receiving the spot price as illustrated in Figure 7–4. The risk they face is that prices will decrease below what they initially paid for the grain. Correspondingly, feed mills are constantly buying grain and processing the grain into livestock feed stocks. Feed mills face the risk that grain prices will increase. A swap dealer could enter a deal with an elevator wherein the elevator would pay the current spot price and in return receive $2.70 per bushel for corn. The dealer would say to the feed mill: pay $2.80 per bushel for corn and receive the current spot price. Figure 7–4 shows these transactions. The elevator has mitigated the risk of declining prices by passing the spot price on to the dealer in return for a stable price of $2.70 per bushel. The feed mill pays $2.80 per bushel for corn while receiving the spot price to actually buy the corn in the spot market. Both feed mill and elevator have managed the risk of price change by entering the swap deal. The swap dealer receives $2.80 per bushel for the corn and pays $2.70 per bushel, thereby earning a 10-cent per bushel margin.

This corn marketing swap would, of course, have a reference price that both elevator and feed mill would view as the spot price that they agree to pay and receive. Additionally, the swap deal would specify the quantity, quality, and time frame for the deal, such as 10,000 bushels per week of #2 yellow corn. This type of commodity works exactly like the natural gas example. The deal is *Pareto optimum* because each participant gains; the elevator and feed mill gain price risk management and the swap dealer earns a positive margin.

Corn Production Swap

Corn production is a discrete operation. Corn is planted in the spring and harvested in the fall and then stored throughout the year until next harvest. Most corn producers do not have a continuous stream of income from corn sales unless they have on-farm storage and pull from

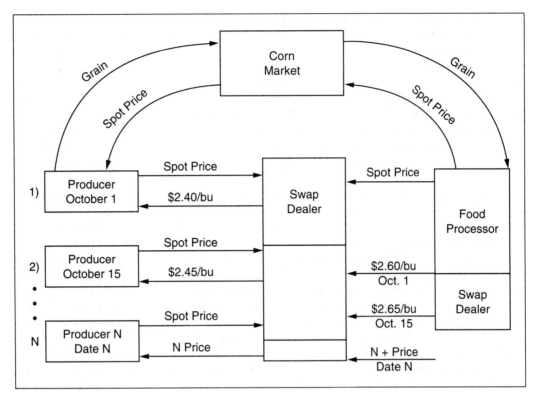

Figure 7–5   Corn production swap example

it on a regular basis and sell into the marketplace. Instead, they either sell at harvest or store and later sell. Some large corn producers produce several types of corn that will mature at different times and thus have several selling opportunities in the fall harvest period, but still the process for most producers is only one or two selling times for the grain. Unlike elevators in the previous example, producers are not attractive targets for traditional commodity swaps.

A swap dealer would have to create a tier or ladder of maturities for producers. Whereas in typical commodity swaps the swap dealer is merely connecting two parties together, a ladder of maturities requires numerous participants on one side of the swap. A swap dealer would enter into a deal with producer X as follows: on October 1, pay the spot price and receive $2.40 per bushel for 100,000 bushels of #2 yellow corn. The dealer would enter into another swap deal with producer Y: on October 15, pay the spot price and receive $2.45 per bushel for 90,000 bushels of #2 yellow corn. The swap dealer would continue building the tier or ladder of maturities until they have a package that would be attractive to a feed mill or food processor. The ladder of maturities would mimic a larger grain merchant such as an elevator. The food processor's swap deal might include the payment of the fixed price for one time period and that same price plus a storage fee for the next time period. Notice in Figure 7–5 that the swap deal is identical to the corn marketing swap (Figure 7–4) in operation except that the left leg of the swap contains more than one participant. Obviously, the right side of the swap deal could be constructed in a similar manner if no single food processor or feed mill was sufficient to cover the left side of the deal. Ladder swaps are sophisticated and complex for the swap dealer, but for the individual corn producer in the example, they are very simple contracts.

Corn producer X's swap deal is very simple. The producer agrees to pass to the swap dealer on October 1 the reference spot price for 100,000 bushels of #2 yellow corn and in turn receive $2.40 per bushel. This is very similar to a typical forward contract in the financial impact to a producer. With a forward contract, the producer agrees to sell the corn at a specified price. So regardless of what the market price is at the time of the contract deal, the producer has to sell the grain to the contracting party. With a swap deal, the producer is free to sell the corn to whomever they wish. They are obligated to pass on a reference price for a given quality and quantity of grain to the swap dealer and in return they will receive a fixed price. The producer may or may not actually sell the grain at that time. A forward contract and a swap contract to a producer are mechanically very different, but similar in financial outcomes.

To illustrate this similarity, consider two different producers. Producer A enters into a forward contract for October 1 for 10,000 bushels and will receive $2.40 per bushel. Producer B enters into a swap deal for October 1 for 10,000 bushels and will receive $2.40 per bushel and pay out a reference spot price. On October 1, the local spot price is $3.00 per bushel. Producer A will have to deliver 10,000 bushels and will receive $2.40 per bushel, and the local price of $3.00 is of no importance to him. Producer B will receive $2.40 per bushel for 10,000 bushels and will have to pay to the swap dealer $3 per bushel. Whereas producer A had to sell the grain for $2.40 per bushel, producer B does not. Producer B could sell the grain and receive the $3 per bushel so he would have the money to pass on to the swap dealer, but he doesn't have to. He could simply pay the $3 per bushel, receive the $2.40 per bushel, and store the grain for later sale. The two producers would have similar financial outcomes only if producer B sells the grain for the current spot price of $3 per bushel and passes it on to the dealer. If producer B does anything else, which he could under the swap deal, the financial outcomes would no doubt be different.

The swap deal gives producer B more flexibility than the forward contract deal for producer A. In the previous example, if the price is higher than the swap transaction, producer B could sell for the higher price and have the income to pass on in the swap deal, or the crop could be held in storage for later sale. Likewise, if the spot price was lower than the swap price, the producer could sell and pass along the income, or hold the commodity for later sale. The swap deal requires producer B to meet the spot price financial obligation, but does not require any physical selling action. How producer B chooses to meet the obligation is his choice. If he meets the spot price obligation out of other income, he still has the product to sell at a later date.

A Final Note on Agricultural Swaps

The authors do not know of any commodity swapping activities in the agricultural industry. The two examples are presented as representative of potential opportunities that exist in the grain and oilseed industry. Livestock swaps are possible as well. The lack of swaps in the agricultural industry may be a result of plenty of other risk management choices. Futures contracts are widely available for most major agricultural commodities and since 1983 most agricultural commodities have options as well. But that argument could be applied to both currency and interest rate swaps also. Currency and interest rate futures have been widely available for the last 20 years as well as options.

Regardless of the reasons why swaps haven't developed in agricultural commodities, the fact remains that they can be very useful tools to help lower costs and provide risk management opportunities, just as they have done for currency, energy, and interest rate trades. Once the swap market started on interest rates, large banks quickly established swap divisions to capture some of the swap margins for themselves. The same could occur in the grains and oilseed markets. Large grain companies could construct commodity marketing swaps as well as ladder swaps for producers. Since all swap deals require a reference price, numerous arbitrage opportunities would be created as well. The potential for agricultural commodity swaps is enormous.

Interest Rate Swaps

Interest rate swaps are by far the most popular form of swaps. They involve swapping the cash flows from fixed rate loans with variable rate loans. Banks usually offer loans to businesses either with a fixed interest rate over the maturity of the loan or a variable rate that is subject to change on fixed intervals (daily, weekly, quarterly) and tied to some common reference rate such as the Treasury Bill (T Bill) rate or the London Interbank Offer Rate (LIBOR). If a business borrows money with a fixed rate, they do not have an interest rate risk. However, they will have to pay a higher rate for the loan initially versus a variable rate loan. Banks generally offer both types of loans as a convenience to customers and as a way to manage their own interest rate risks. Banks will change the relative rates between a fixed and variable rate loan depending upon their own needs. For example, if a bank needed more variable rate loans in its loan portfolio it would price the fixed rate loans relatively higher than variable rate loans as an inducement to customers to take more variable rate loans, and vice versa. Banks also charge different interest rates to different customers based on each customer's credit rating.

Therefore, within each bank is the potential for several different interest rates as a result of the bank's own risk management strategies and each customer's credit worthiness. Multiply the multitude of different loan rates within each bank by the number of banks within the U.S. (approximately 6,000) and the opportunity for swapping interest rate cash flows is almost limitless.

Fixed versus Variable Loan Swap Example

Bank A offers a fixed rate loan of 10 percent or a variable rate loan of the LIBOR rate (LIBOR currently at 3 percent) plus 4 percent to counterparty A. Counterparty A wants a variable rate loan. Bank B offers a fixed rate loan of 11 percent or a variable rate loan of the LIBOR rate plus 3 percent to counterparty B. Counterparty B wants a fixed rate loan. The situation can be enhanced by a swap dealer because the necessary elements are in place—two businesses that face different loan rates from their individual banks and want different types of loans. A swap dealer could offer to counterparty A the following deal: pay the LIBOR rate plus 3.5 percent to the swap dealer, and in return receive a 10 percent cash flow. The dealer would offer to counterparty B: pay a fixed cash flow to the dealer of 10.5 percent and in return receive the LIBOR rate plus 3 percent. Counterparty A then borrows fixed from bank A and counterparty B borrows variable from bank B, as shown in Figure 7–6.

Counterparty A has a cash flow obligation to bank A of 10 percent which they receive from the swap dealer. Counterparty A then pays to the dealer the LIBOR rate plus 3.5 percent. In effect counterparty A has the cash flow obligations of a variable rate loan to the dealer, but the legal obligation to the bank of a fixed rate loan. The financial result to counterparty A, however, is a variable rate loan that is the LIBOR rate plus 3.5 percent which is 50 basis points less than they would have had to pay to bank A if they had borrowed a variable rate loan. Counterparty B has a cash flow obligation to the dealer of providing a 10.5 percent fixed rate and in return they receive from the dealer the LIBOR rate plus 3 percent which covers their loan obligation to bank B. Counterparty B has the cash

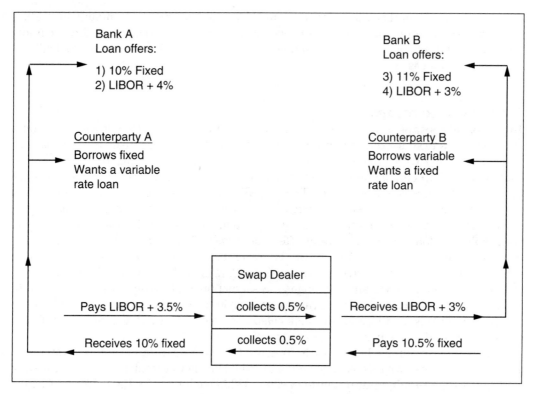

Figure 7–6   Fixed versus a variable rate loan swap

flow obligations of a fixed rate loan to the dealer, but the legal obligation to bank B of a variable rate loan.

Counterparties A and B as well as the dealer are all better off by trading with each other than if they had not traded. The deal is said to be *Pareto optimal*. Counterparty A wanted a variable rate loan, but the best they could get from their bank was LIBOR plus 4 percent. The swap deal netted them a LIBOR plus 3.5 percent, or a gain of 50 basis points. Counterparty B wanted a fixed rate loan, but the best they could get from their bank was 11 percent. The swap deal netted them a 10.5 percent, or a gain of 50 basis points. The swap dealer receives from Counterparty A the LIBOR rate plus 3.5 percent and passes on to Counterparty B the LIBOR rate plus 3 percent, for a net gain on that leg of the swap 50 basis points. The swap dealer receives from Counterparty B a fixed cash flow of 10.5 percent and passes on to Counterparty A 10 percent, for a net gain on that leg of 50 basis points. The swap dealer earned 50 basis points on each leg for a net gain of 100 basis points. All three are better off financially by swapping and all counterparties received the type of financial obligation that they wanted.

The swap dealer, once both legs of the swap are put into place, has no rate change risk. Counterparty A pays the LIBOR rate plus 3.5 percent and the dealer passes on only the LIBOR rate plus 3 percent to Counterparty B who in turn pays bank B the LIBOR rate plus 3 percent due on the variable rate loan. The swap dealer receives from Counterparty B a fixed rate cash flow of 10.5 percent and passes on only 10 percent to Counterparty A who in turn pays bank A the 10 percent fixed rate due on the loan. Counterparty B also has no rate risk. The only rate risk in the example is the change in the LIBOR rate. Counterparty B is receiving the LIBOR rate plus 3 percent; that will pay off the cash flow needs of the loan to bank B and therefore has mitigated the variable rate loan risk. Counterparty B is paying to the dealer a fixed rate and thus has no overall rate risk. Counterparty A, on the other hand, does have a rate risk. Counterparty A is borrowing fixed and is receiving the necessary funds to pay the loan cash flow from the swap dealer. However, they have to pay to the dealer the LIBOR rate plus 3.5 percent. If the LIBOR rate increases, they will have to pay more to the dealer each pay period. Counterparty A has assumed the risk of increasing interest rates by entering the swap deal. Counterparty A has received the benefit of a cheaper variable cash flow of 50 basis points by entering the swap deal than if they had taken a variable rate loan with bank A, but nonetheless still has the risk of the LIBOR rate going up. Accordingly, the swap deal, while Pareto optimal for all three, mitigates rate risk for only two of the three participants.

A Closer Look at Counterparty A

Counterparty A in the previous example could have borrowed a variable rate loan from bank A for a rate of the LIBOR rate plus 4 percent. If Counterparty A had done that, then they would face the risk of the LIBOR rate going up. The swap deal did nothing to handle the risk of rate change for Counterparty A; it only reduced by 50 basis points the cost of the loan. Counterparty A has a fixed rate loan with bank A and a swap contract with the swap dealer that has reduced the cost of a variable rate loan, but the risk that LIBOR rates will change is still viable. Counterparty A must either absorb the risk or manage the risk. Counterparty A can manage the LIBOR rate risk by hedging with either futures or options contracts. If Counterparty A decides to hedge the risk of LIBOR rate changes after they have also entered into a swap deal, then they have created a **complex derivative.** If Counterparty A had not entered the swap deal and merely hedged the risk of LIBOR rate changes, the hedge would have been a simple futures or options hedge. However, by entering the swap deal, the LIBOR rate change risk is now part of the swap contract and now a futures or options hedge is contingent upon the swap contract and becomes a more complicated situation, thus the term complex derivative. Complex derivatives are fully covered in Chapter 8, including a situation similar to the risk that Counterparty A has in this example.

Currency Swaps

Currency swaps generally involve the swapping of fixed rate loans denominated in one foreign currency with another fixed rate loan in another currency. To be sure, currency swaps do involve fixed rate loans denominated in one currency for a variable rate loan denominated in another currency, but these are currency plays that are generally used by extremely large operations and are fairly complex. Currency swaps differ from most commodity and interest rate swaps in that they almost always involve notional swaps.

   The example in Table 7–1 involves a Canadian firm (CF) and a U.S. firm (USF). The CF has 15,000,000 Canadian dollars (CD) that it will swap with USF for their 9,000,000 United States dollars (USD). This assumes that the Canadian firm needs USD and the United States firm needs CD. The tenor of the swap is 3 years with interest rates in Canada at 6 percent and in the United States at 5 percent. The Canadian firm gets 9,000,000 USD for three years and pays an annual interest payment of 450,000 USD ($9,000,000 at 5 percent). The United States firm gets 15,000,000 CD for three years and

Table 7–1   Currency Swap Fixed Rate Loan for Fixed Rate Loan

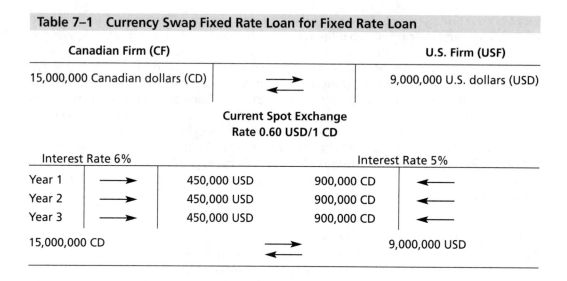

pays an annual interest payment of 900,000 CD. At the end of the three years, the two firms swap their notional principals.

The two firms have set the exchange rate at the current spot rate of 0.60USD/1CD for the initial swap of the notional principals and for the swap back in three years. However, each firm is subject to currency rate risk for each year's interest payment as each has to pay in the other's currency. Since the yearly interest payments are significantly lower than the notional principals, each firm has reduced their currency rate risk. They have, in essence, swapped the larger principal risk for a much smaller yearly payment currency risk. These yearly payments could, of course, be hedged with futures or options.

One of the counterparties could have its loan as a fixed rate loan and the other as a variable rate, both could be fixed as in the example in Table 7–1, or both variable. The mechanics will be similar to the example in Table 7–1, except the interest payments might involve a variable loan such that each year's interest payment would be an unknown.

What did each firm gain? The only certain gain was a known currency exchange rate for the initial and final swap of the notional principals. Whether or not this was a financial gain over what could have been attained in the foreign currency markets is unknowable.

A Final Word on Swaps

The swap market is a very large market and one without any major governmental oversight. Exchange-traded futures and options are large, open public markets with oversight by the Commodity Trading Commission. Additionally, each exchange has separate clearing corporations that insure that defaults are quickly compensated to the other party, thus the fact that futures and options have essentially no default risk built into the price. Swap markets have none of these provisions and will not likely get any in the foreseeable future. Thus swaps carry default risk and the creditworthiness of each counterparty has to be assessed.

On the other hand, because swaps are not exchange traded they enjoy a measure of privacy in contract dealings. Swap contracts lack liquidity since they cannot be retraded, however they can also be negotiated for unique quantities, qualities, and contract provisions.

CHAPTER 7—QUESTIONS

1. How can a swap both manage the risk of price change and have the ability to actually change the price level for trade?

2. What would cause a business to accept an interest rate swap deal that moves the business from having a known fixed rate on a loan to the equivalent of a variable rate loan? Doesn't that move the business from having no rate risk to a rate sensitive environment?

3. Counterparty A can get a fixed rate loan from her bank for 10 percent and a variable rate loan at a rate of LIBOR plus 3 percent (LIBOR currently at 3 percent). Counterparty A is willing to borrow fixed, but really wants a variable rate loan because of the much lower cost. What kind of a swap deal would appeal to counterparty A? Why?

4. What is the importance of a reference price for commodity swaps?

5. Why is the concept of Pareto optimum important to swap deals? Explain.

Complex Derivatives

This chapter examines the process to handle multiple risks using complex derivatives and also discusses how to evaluate and mitigate cloistered risks.

OVERVIEW

Complex derivatives became very popular during the decade of the 1990s, and by the early part of the twenty-first century were feared as well. The California energy crisis and the collapse of the giant energy company Enron fueled the public's fear of derivatives because they were alleged to be at the heart of both problems. Derivatives, especially complex ones, are very powerful risk management tools, yet they have to be fully understood and used correctly or they can be equally powerful in causing financial harm. This chapter will outline the process and use of complex derivatives.

Introduction

During the last two decades of the twentieth century as computers became more powerful and available to the average individual, more and more financial analyses were being performed than at any other comparable time in history. Analysis that took the capacity of entire computer centers at universities in the late 1970s could now be performed on a simple desktop computer in anyone's home or office. Naturally, financial analysis and the use of futures and options became more complex. During the late 1980s futures, options, swaps, and other types of contracts started being referred to as derivatives. Part of the reason for the change in name was due to the sheer arrogance in the industry—why refer to a simple out-of-the-money option when the words "double complex derivative" could be used? However, a broader term was needed to relate to the financial industry that once contained only futures contracts, but now also had options, swaps, and other contracts. The term *derivative* fit nicely and quickly gained wide acceptance.

The Role of Multiple Risks

Most of the financial risks that agricultural producers and agribusinesses face can be managed with simple hedges with futures or options. An Iowa corn farmer needs price risk protection during the growing season on corn, and a simple futures or options hedge fits the bill adequately. A large food processor that has a variable rate operating loan needs rate protection during the loan duration, and a simple futures or options hedge will do the job just fine. Yet even

What are the risks?
  1) Corn prices declining during growing season
  2) Soybean prices declining during growing season

Can the risk be mitigated?
    Yes

How can the risks be mitigated?
    Futures and options contracts exist on both commodities as well as forward contracts.

Figure 8–1   Partial risk and mitigation profile for a typical
Midwest farm with multiple single risks

the simplest of businesses will often have more than one risk and sometimes one risk is embedded within another. Risks that are fairly easy to separate, and keep separated, are referred to as **single risks.** Risks that are embedded within another risk and are therefore not easy to separate or keep separated are called embedded or **cloistered risks.**

A Multiple Single Risk Farming Example

A Midwest corn farmer is generally a soybean farmer as well for agronomic reasons. Thus, a simple farming operation in the corn belt states has multiple single risks—corn and soybean price movements. Going back to Chapter 2 and picking up the **Risk and Mitigation Profile (RMP),** the first step in the RMP is to list the financial risks. A typical farm will have several financial risks, but let's confine ourselves to the two price risks for now. A corn and soybean farmer faces the risk that during the growing season the price of corn will go down and likewise for soybeans. A simple RMP for a Midwest farm is illustrated in Figure 8–1. Only the first three steps are illustrated for now to emphasize the nature of multiple single risks.

The simple farming example has two risks: corn and soybean price changes that are easy to separate and keep separated. If the farmer views these risks independently, then a decision to simply hedge with futures contracts might be adequate to manage the price risks.

A Farm with Cloistered Risks Example

All too often, risks are not easily separated. And for those that can be separated, it is often very difficult to keep them separate. A cloister is a covered passage and thus is an ideal term for one risk that leads to another in a way that implies a linkage or cover. Consider the previous example of a simple farm in the Midwest that grows both corn and soybeans. The partial RMP in Figure 8–1 identifies the two risks as separate. However, most farms also have to have an operating loan to produce the crop and many will also have a land mortgage that must be paid as well. Assume that both the operating loan and land loan are fixed rates and thus do not have the risk of rate change. Both the operating loan and the land loan are directly tied to the production of corn and soybeans on the farm. The land loan was to buy the initial production unit so that corn and soybeans could be produced and the operating loan provided the financing for inputs.

The risk of price changes for corn and soybeans is embedded in the loans; stated another way, the loans have cloistered risks. Indeed, prudent loan officers will always look at the repayment capacity of the borrower via the borrower's collateral and other sources of

What are the risks?
    Corn and soybean prices insufficient to cover operating and land loan payments

Can the risk be mitigated?
    Maybe

How can the risks be mitigated?
      Futures and options contracts exist on both commodities as well as forward contracts—but can they be used to generate prices sufficient to cover the loan risks?

Figure 8–2   Partial risk and mitigation profile for a typical
Midwest farm with cloistered risks

funds, but also by the ability of the business activity to generate sufficient income to repay the loan. Banks know that most business loans have multiple risks and those risks are cloistered. Here the word derivative is also applicable. Part of the risks of the loans are derived from the risks of the price changes on the corn and soybeans.

With derived or cloistered risks, the farmer's situation may need to be handled differently. Figure 8–2 exhibits the partial RMP for the farm with cloistered risks. As the risks become cloistered, the ability to mitigate the risks becomes more uncertain. The loans have an exact cost to the farmer, and in order to generate enough income from the sale of corn and soybeans, the crops need to sell for a certain price. The price objective may or may not be possible via futures, options, or forward contract.

If the price objective via risk management tools is unattainable, the farmer faces the decision to either use one of the tools and fall short of the objective or absorb the price risk and face the potential for financial damage or gains. For example, the farmer determines that to fully cover both the operating and land loan, his corn crop will have to sell for at least $2.00 per bushel and his soybeans will have to bring in at least $5.55 per bushel. The farmer must now determine if there is a hedging opportunity with either options or futures that would cover those pricing points. Numerous strategies can be constructed to help the farmer manage the embedded risks. Figure 8–3 shows the complete RMP with a potential strategy for managing the risks.

The farmer can earn a positive income above loan payments with a fair degree of certainty by hedging the corn and soybean crops. Should the farmer do this? Although the RMP is fairly complete, it could provide more help for the farmer to make the decision. For example, the price objectives are stated in terms of loan repayment requirements, and not the actual cost of production for the two crops. A more complete price objective that contained all production costs would provide the farmer with better information and would be helpful for the farmer in making longer term decisions. Furthermore, alternative risk management tools such as options and forward contracts could provide more pricing opportunities for the farmer. And lastly, what are the likely consequences if the farmer decides to not hedge?

RMPs provide a framework to help understand multiple risks and how they might be handled and what the likely consequences will be using various risk management strategies. To be sure, no profile will ever be complete, but they do afford a structured look at multiple risks and how those risks can be managed. There is an old military saying—"No battle plan

What are the risks?

Corn and soybean prices insufficient to cover operating and land loan payments

Can the risk be mitigated?

Maybe

How can the risks be mitigated?

Futures and options contracts exist on both commodities as well as forward contracts—but can they be used to generate prices sufficient to cover the loan risks?

Cost and benefits?

The farmer knows that the probability of producing a corn crop that is 125 bushels per acre or more is 0.9 and he is comfortable with that level. Likewise, for soybeans, the 0.9 probability is for 45 bushels or more. His property is equally divided with 800 acres of each crop.

$800 \times 125 = 100,000$ bushels of estimated corn production

$800 \times 45 = 36,000$ bushels of estimated soybean production

The operating loan amount due including interest is $200,000 and likewise for the land payment.

$400,000 loan service necessary

The farmer allocates $1/2$ of the loan payment to each crop since they are equal in acreage.

$200,000 ÷ 100,000 bushels of corn = $2.00 per bushel for corn

$200,000 ÷ 36,000 bushels of soybeans = $5.55 per bushel for soybeans

December corn futures @	$2.45/bushel
Localized basis	– .15/bushel
Costs of hedging	.05/bushel
Estimated net hedged price	$2.25/bushel
November soybeans @	$5.70/bushel
Localized basis	– .10/bushel
Costs of hedging	– .05/bushel
Estimated net hedged price	$5.55/bushel

Estimated hedged income

$100,000 \times \$2.25 = \$225,000$

$36,000 \times \$5.55 = \$199,800$

$\$424,800$

Figure 8–3   Risk and mitigation profile for a typical
Midwest farm with cloistered risks

Do I want to mitigate the risks?

The farmer has a debt obligation of $400,000 and an estimated hedged income of $424,800.

The loan debt is a certain number. The estimated income is conservative given the 0.9 probability of crop yields. Localized basis values are fairly stable and the cost of hedging would be stable as well. Income would increase if yields increased above the estimated levels. The farmer feels the estimated hedged income is a fairly solid number. If he hedges, he will earn approximately $25,000 above paying off the operating loan and making his annual land payment.

Figure 8–3 (*continued*)

survives contact with the enemy." Yet generals still plan battles because to not plan would, in their minds, be unthinkable. Price risk management is no different. No RMP will ever be correct, but to not develop one, especially where embedded risks are concerned, is unthinkable.

Financial Engineering

As more and better financial information becomes available, particularly in conjunction with computer analysis, more sophisticated risk management tools have evolved. During the movement toward more financial analysis, the term **financial engineering** has emerged. To engineer means to plan and execute and no doubt many operations, including finance, could appropriately be referred to as having been engineered. Much to the chagrin of most engineers, however, the term engineering is applied to many odd and undeserving enterprises. The authors feel likewise and for that reason will avoid the term financial engineering in favor of the old but more appropriate term, **financial management.** To be sure, it is important to develop the appropriate risk management tool to handle the identified risk, yet equally important is the not so exact process of managing the risk and tool simultaneously.

Most of the spectacular failures of companies that have used derivatives during the last few years are not a result of the derivative itself being constructed wrong, but rather the failure of management to properly apply and monitor the tool. Thus, the engineering (designing and constructing) of derivatives is certainly important, but managing the risk is paramount. Complex derivatives have two parts: designing and constructing and management.

Designing and Constructing Complex Derivatives

Proper design and construction of anything has to start with a purpose. If a home design starts with the needs of the family first, the final home that is constructed will likely satisfy those needs. If a family is forced to move into a home, sight unseen, then there is a better-than-even chance that the home will not meet the needs of the family. Most families will not have a home designed and built specifically for them, but they do shop for homes that are already designed and built in an effort to find one that meets their needs. Designing and constructing complex derivatives is very similar to the home example. A unique complex derivative can be designed and constructed to meet a specific risk, or management can take a specific risk and shop for a derivative that will mitigate the risk. Either way is fine because they both start with the risk to manage in mind—custom build a derivative or buy one off the shelf to handle the risk.

The biggest problem in complex derivatives is the derivative in search of a home. Many firms have packaged complex derivatives that they starkly market to anyone who will buy them, regardless of whether or not they properly meet the risk needs of the buyer. Undeniably simple futures and options contracts are readily sold to anyone who wants to buy them, regardless of whether or not they meet the needs of the buyer—but they are simple derivatives. Complex derivatives require greater effort to fully understand how they function and if they are not appropriate for the risk at hand, they may in fact increase—not mitigate—the risk. In short, the best solution is to start with the RMP, which defines and outlines the risks that the business faces, and not start with the derivative regardless of whether it is simple or complex.

A Quick Example of Abuse

In Chapter 6 (which covers options), an example was presented of a synthetic futures hedge. If a business needs to short hedge with futures, an alternative hedge is proposed by brokers called a *synthetic futures hedge* that involves the simultaneous buying of a put and the selling of a call at the same at-the-money strike prices. Brokers will promote this as a "zero cost hedge" and explain that the premium that would be required for buying the put is offset by the selling of a call, thus zero cost. As the example in Chapter 5 shows, the synthetic futures hedge does indeed mimic a futures hedge and would be considered theoretically a hedge. The simultaneous buying of a put and the selling of a call is a complex derivative. This is an example of two simple derivatives that are packaged together and called by brokers who promote them either "zero cost" hedges or "synthetic" hedges. They are a complex derivative in search of a buyer, and because they mimic a futures hedge they provide risk management and are generally ignored by serious hedgers. They remain in wide use, not because they provide proper risk management, but because brokers get two commissions instead of one with a simple derivative. Zero cost hedges were designed and constructed to meet the needs of the seller, not the business that has the risk.

In fairness, the simultaneous buying of a put and the selling of call can be a strategy that businesses use to provide some measure of risk management and speculate to earn additional income. The process is generally known as a synthetic forward and has almost unlimited **combinations** of different strike prices. Proper use of this type of complex derivative requires a detailed trading plan and full knowledge of the types of risks the business is willing to accept.

Combinations

Complex derivatives are used to cover either multiple single risks or cloistered risks. When several single risks are bundled together and risk management tools applied to mitigate the risks, the result is a complex derivative that is generally known as a **stacked derivative.** Stacked derivatives often occur over time. A business manages one risk only to have another one appear and then another risk management tool is used for that one, and so on, until the business has essentially stacked one tool on top of another. The end result is a complex set of derivatives all piled on top on one another. And while this often occurs, the idea behind a stacked complex derivative is a more systematic approach to the risks the business faces and how to properly manage those risks. The term stacked derivative will be used for multiple single risks and the term **cloistered derivative** will be used for embedded or cloistered risks. Both terms involve combinations of various risk management tools. The best tool to start the process is the RMP.

Stacked Derivative Example

A corn and soybean farmer in the Midwest has both crops planted and both are in the early stages of growth. It is May 15 and the farmer wants to manage the risk that the prices of both crops will go down from now until harvest time in the fall. Figure 8–4 exhibits the RMP for the farmer. In the example in Figure 8–4, the raw basics are outlined and the farmer is left with total uncertainty on which tool to use. Figure 8–5 takes the Costs and Benefits section of the RMP and adds standard risk analysis to help the farmer construct his stacked derivative.

What are the risks?

    Corn and soybean prices decrease

Can the risk be mitigated?

    Yes

How can the risks be mitigated?

    Futures and options contracts
    Forward contracts available from local elevator

Cost and benefits?

    The farmer is very risk averse and thus uses his historical low on production levels because he feels a drought may occur this season.

    Corn production estimated at 87.5 bushels/acre
    Soybean production estimated at 25 bushels/acre

    The farmer has 400 acres of corn and 400 acres of soybeans planted.

Estimated corn production	$400 \times 87.5 = 35,000$ bushels
Estimated soybean production	$400 \times 25 = 10,000$ bushels

    The farmer estimates his total cost of production as: corn $2.02/bushel and $5.95/bushel for soybeans.

Current December corn price	$2.40/bushel
Estimated local basis	− .20/bushel
Cost of hedge	− .05/bushel
Estimated target price	$2.15/bushel
Current November soybean price	$6.03/bushel
Estimated local basis	− .10/bushel
Cost of hedge	− .05/bushel
Estimated target price	$5.88/bushel

Options are unavailable that far out

July corn $2.40 put is trading for 9 cents/bushel

July soybean $6.00 put is trading for 11 cents/bushel

    The farmer would have to roll the option hedges in late June or offset and remain unhedged until harvest.

    Forward contract offers: $2.10/bushel for corn; no soybean contract offer

Do I want to mitigate the risks?

    Yes, but how?

Figure 8–4   Risk and Mitigation Profile for corn and soybean farmer

Costs and benefits?

Futures:

Estimated corn futures target price $2.15

Estimated local basis standard deviation = $0.08
    Local basis range for one standard deviation
        $0.12 – $0.28

        Estimated corn target price range $2.07–$2.23

Estimated soybean futures target price $5.88

Estimated local basis standard deviation = $0.05
    Local basis range for one standard deviation
        $0.05 – $0.15

        Estimated soybean target price range $5.83–$5.93

Options:

July options will have to be offset in late June, so the farmer checks his moving average price charts and sees no clear bullish or bearish signals. He is reluctant to use options at this point.

    Forward contract offer   Corn $2.10/bushel
                                           Soybeans none

Corn:

1)   Hedged with Futures:

      Corn price range = $2.07–$2.23
      Cost of production = $2.02
         Estimated margin $0.05–$0.21
         Estimated gross 35,000 × 0.05 = $1,750   35,000 × 0.21 = $7,350
            Range = $1,750 – $7,350

      Degree of certainty: moderately high

2)   Forward Contract: $2.10 – $2.02 = $0.08 margin
        Gross = $2,800

      Degree of certainty: high

Soybeans:

1)   Hedged with futures:
        Soybean price range = $5.83–$5.93

        Cost of production = $5.95

Figure 8–5   Costs and benefits of stacked derivatives

Estimated margin $(0.12)–(0.02)

Estimated gross   10,000 × –0.12 – $(1,200)

10,000 × –0.02 = $(200)

Range = $(200)–$(1,200)

Degree of certainty: moderately high

2) Forward contract: not available

Stacked: Range $1,750 + (1,200) = $550

with futures $7,350 + (200) = $7,150

Stacked forward: $2,800 + (200) = $2,600

$2,800 + (1,200) = $1,600

Figure 8–5   (*continued*)

The farmer looks more closely at the factors that have the most variation on the estimated futures target price—the estimated ending basis. If the distribution of the ending basis is normal, one standard deviation would pick up 67 percent of the likely values. For corn, the estimated range is from 5 cents to 21 cents per bushel and thus gives the farmer a better idea of how risky the estimated target price really is (likewise for soybeans). The end result is that if the farmer stacks two futures hedges together, his range of potential gross income is $550 to $7,150 and with a forward contract/soybean futures stack the range is $1,600 to $2,600.

If the farmer opts for the forward/futures stack, his estimated gross income is positive within a range of $1,600 to $2,600. However, the futures/futures stack will still leave the farmer with a positive estimated gross, albeit only $550, but the upper range is $7,150.

Notice how two very simple risks that are easy to separate and keep separated, when viewed as a total package for the farmer, become intertwined in the gross farm income. When the two are stacked together the farmer gets a broader picture of the farm's likely outcome.

Cloistered Derivative Example

Cloistered or embedded risks are best viewed as "if, then" problems. If a certain risk is assumed or created, then another risk occurs. A grain company that sells corn has the risk that corn prices will decrease, and if the corn is sold for delivery and payment in another currency it has the risk that the foreign exchange will be unfavorable. A decision to feed cattle creates the risk that cattle prices will decrease during the feeding period and additionally that the price of feed will increase. One risk is embedded in or cloistered by another. One of the best examples of a cloistered risk is that of animal feeding operations. Beef, pork, and poultry are finished with higher quality rations for the last period before slaughter. Both the price of the finished animal and the cost of the feed are variable and risky. The poultry industry has moved to almost total vertical integration as a means to handle the risk, and pork is doing likewise. The beef industry still has the most exposure. As an example, consider the process of putting steers on feed for a 120-day feeding period consisting of a ration composed mostly of corn and soybean meal. The feeding operation will transform the cattle from feeders to slaughter-ready fed cattle via a concentrated feeding ration composed of a grain (corn) and a concentrated protein (soybean meal). This feeding process has triple price risks—fed cattle, corn, and soybean meal with each embedded or cloistered by the others. Figure 8–6 displays the RMP for the cattle feeder.

What are the risks?

1) Fed cattle prices declining over the 120-day feeding period
2) Corn prices increasing over the 120-day feeding period
3) Soybean meal prices increasing over the 120-day feeding period

Can the risks be mitigated?

Yes

How can the risks be mitigated?

Futures and options contracts exist on all three commodities
No formal contracting opportunities exist

Costs and benefits?

The cattle feeder estimates he will have approximately 200,000 lb of fed steers available for sale in 120 days. He fills his storage bins with corn today for $2.20/bushel and soybean meal at $160/ton. He will need to purchase 15,000 bushels of corn and 100 tons of soybean meal in 2 months.

Futures:

October live cattle futures	$70.00/cwt
Estimated local basis	−$0.50/cwt
Cost of hedge	−$0.10/cwt
Estimated target price	$69.40/cwt

Standard deviation of basis = $0.40
Range of estimated target prices = $69.00 – $69.80

July corn futures	$2.20/bushel
Estimated local basis	−$0.15/bushel
Cost of hedge	+$0.05/bushel
Estimated target price	$2.10/bushel

Figure 8–6   Risk and mitigation profile for a cattle feeding
operation with cloistered risks

July soybean meal futures	$160.00/ton
Estimated local basis	−$5.00/ton
Cost of hedge	+$0.30/ton
Estimated target price	$155.30/ton

Standard deviation of basis = $2
Range of estimated target prices = $153.30 − $157.30

Options:

October live cattle at-the-money put option

Strike price	$70.00/cwt
Premium	$2.70/cwt
Cost of hedge	$0.06/cwt
Estimated option target price	$67.24/cwt

One-step on-the-money put option

Strike price	$71.00/cwt
Premium	$3.65/cwt
Cost of hedge	$0.07/cwt
Estimated option target price	$67.28/cwt

One-step out of the money $69 strike price, $1.75 premium

Strike price	$69.00/cwt
Premium	$1.75/cwt
Cost of hedge	$0.05/cwt
Estimated option target price	$67.20/cwt

July corn at-the-money call option

Strike price	$2.20/bushel
Premium	$0.05/bushel
Cost of hedge	$0.03/bushel
Estimated option target price	$2.28/bushel

One-step in the money

Strike price	$2.10/bushel
Premium	$0.14/bushel
Cost of hedge	$0.04/bushel
Estimated option target price	$2.28/bushel

Figure 8–6   (*continued*)

One-step out of the money

Strike price	$2.30/bushel
Premium	$0.03/bushel
Cost of hedge	$0.02/bushel
Estimated option target price	$2.35/bushel

July soybean meal at-the-money call option

Strike price	$160.00/ton
Premium	$5.00/ton
Cost of hedge	$0.25/ton
Estimated option target price	$165.25/ton

One-step in-the-money call option

Strike price	$155.00/ton
Premium	$8.00/ton
Cost of hedge	$0.30/ton
Estimated option target price	$163.30/ton

One-step out-of-the-money call option

Strike price	$165.00/ton
Premium	$4.00/ton
Cost of hedge	$0.22/ton
Estimated option target price	$169.22/ton

Do I want to mitigate the risks?

Yes, but how?

Figure 8–6   (*continued*)

If the cattle feeder does not manage the risk of price change for all three risks together, the end is indeterminate. If, for example, the feeder estimates a breakeven price for the fed cattle at $65 per hundredweight using corn priced at $2.20 per bushel and soybean meal at $160 per ton, then all three prices have to be managed to get the desired results. If corn prices are left to chance, then the breakeven price is likewise at risk. All three are embedded and must be managed simultaneously. Using the information from the RMP in Figure 8–6, the feeder has several choices. If the feeder hedges all three with futures, the worst case price for corn would be $2.19 per bushel and for soybean meal $157.30 per ton, both below the assumed price levels for breakeven using one standard deviation as the risk level. The fed cattle worst price would be $69 per hundredweight, well above the $65 per hundredweight breakeven. If the feeder fully uses a cloistered futures hedge, assuming one standard deviation as an acceptable risk level, the feeding operation will turn a profit.

If the feeder uses options, then a fed cattle price floor can be established at $67.28 per hundredweight using the one-step in-the-money option. The feeder will receive as a minimum $67.28 per hundredweight, regardless of how low cattle prices drop, and if prices increase, the feeder could receive a higher price. A maximum corn price ceiling can be established at $2.28 per bushel using the one-step in-the-money option and a soybean ceiling of $163.30 per ton with one-step in-the-money options. The fed cattle option is above the breakeven price, but the corn and soybean options are not. The extra 8 cents per bushel corn price translates to an additional $1,200 added cost (15,000 bushels × $0.08) and the extra $3.30 per ton for the soybean meals adds $330 (100 tons × $3.30). The option ceiling adds $1,530 total to 200,000 pounds of fed cattle, or an additional $0.765 cent per hundredweight to the fed cattle breakeven of $65 per hundredweight, or a new breakeven with option hedge ceilings on corn and soybeans of $65.765. Since the option hedge on fed cattle establishes a price floor of $67.28, a full cloistered hedge with options would create a profit for the feeder, with price protection and the chance for even greater profit if either fed cattle prices increased or corn or soybean prices decreased.

Of course the feeder needs to select a combination of futures and options to construct the cloistered hedge, depending upon the feeder's objectives, tolerance for price risk, and ability or interest in price forecasting. With the options hedge, price floors and ceilings are created with the opportunity to gain if prices move in the desired direction. Options hedges are, therefore, perfect companions to price forecasting tools. If, for example, the cattle feeder was using a price forecasting service that had an outlook during the next 120 days for generally declining cattle prices and slightly decreasing corn and soybean prices, the feeder may want to use a mixture of futures and options hedges. If cattle prices are forecast to decrease, a futures hedge provides a higher price than the options hedge which would be more beneficial than a futures hedge only if cattle prices increased. The feeder would like corn and soybean prices to be as low as possible; if the forecast is for slightly lower prices, then an options hedge would offer the opportunity to capitalize on lower prices whereas a futures hedge cannot. However, the lowest ceiling corn and soybean options add an extra 76.5 cents per hundredweight to the breakeven cattle price. The feeder now must make a decision concerning how strongly he feels about the possibility of corn and soybean meal price declines. Under this scenario, the feeder may decide to have a cloistered hedge that is a combination of a short futures hedge for fed cattle and one-step in-the-money long call options for corn and soybean meal. The reader can immediately see that the possible combinations for the cloistered hedge in this rather simple example are quite large and all involve the feeder determining their attitude toward various risk levels.

The Power of the Risk and Mitigation Profile (RMP)

Once a good RMP is developed for a particular situation, the user can start adding or subtracting levels of sophistication. Depending upon the user's attitude about price forecasting, numerous price outlook scenarios, as well as an almost unlimited array of option strike prices and premiums, can be incorporated or removed. The same is true for futures hedges. In the previous example, only one standard deviation was used to provide a range of estimated target prices, when two or more would provide added levels of risk confidence. Furthermore, the assumption that basis values are normal distributions may not be true, and other more sophisticated statistical tools may be warranted, depending upon the desire and need of the user. Options also can become more complex as estimates of deltas and pricing formulas are used to add finesse and sensitivity. This is the domain of financial engineers (or as the authors prefer, financial managers) as they build risk management strategies.

Regardless of the level of sophistication the user desires, the RMP format provides the structure to look at the tools *in conjunction with* the identified risks. The RMP model keeps the heart of the problem—price risks—in line with the proposed tools to manage the risk. All too often the original price risk gets forgotten as sophisticated tools emerge and a bigger risk develops—that the original risk is not mitigated by the proposed tool and in fact the proposed tool *adds* risk. Now the user has the original risk and the risk of an improperly applied futures, option, or swap contract. He is under the illusion that the risk is covered (i.e., he is hedged), when in fact he is not hedged and now has additional risk exposure.

Unwinding

If a user has constructed an RMP for a situation, he now has a tool to look at proposed risk management packages offered by either brokers or financial managers. Once a proposed strategy is offered, the important question is simply, Does the contrived scheme mitigate the risk? If not, why not? And if so, how does the strategy compare to the listed costs and benefits of other strategies? Let's look at a couple of examples of using the RMP to unwind (disassemble) a risk management package.

Risk Management Package #1

A marketing firm is asked by the cattle feeder in the previous example in Figure 8–6 to submit a plan to handle the cloistered risks of the feeding operation. The marketing firm's recommendation is listed in Figure 8–7. Using the RMP from Figure 8–6, the cattle feeder needed at least $65 per hundredweight for the feeder cattle, $2.20 per bushel for the corn, and $160 per ton for the soybean meal. The marketing strategy offered in Figure 8–7 does mitigate the feeder cattle price risk since it involves the buying of a put; however, it does not meet the minimum price requirements for the cattle feeder. If the feeder cattle price forecast is stable to slightly bearish, then why hedge the feeder cattle risk with an option that doesn't even cover the breakeven price? A futures hedge would provide an estimated target price with a fairly high probability of occurring that does cover the breakeven price. An options hedge for feeder cattle in this example makes more sense if the price forecast was bullish so the feeder would have a hedged price floor but be able to benefit from a general price increase; likewise for the corn and soybean meal options hedges. The buying of call options mitigates the risk of increasing prices, but if the forecast is for a slightly bullish price move, then why options? A futures hedge would meet the minimum price requirements of the feeder, but the proposed options hedges do not. If the price forecast was slightly bearish, an options hedge would allow for the potential for cash market gains with a price ceiling for the hedge.

This package, while technically correct as a hedge, does not take full advantage of other information. Options hedges are better than futures hedges when prices move in favor of the cash position, and for all three commodities in this example; the forecast was just the opposite. This is not an incorrect recommendation, but certainly doesn't fulfill the feeder's needs nor does it take advantage of other information, such as price forecasting.

Risk Management Package #2

The cattle feeder in the example in Figure 8–6 has his broker submit a hedging plan for the feeding program. Figure 8–8 has the broker's recommendation. Using the RMP from Figure 8–6, the broker is offering a hedge for the feeder cattle, although the options hedge does not

The 3-, 9-, and 18-day moving average for the August feeder cattle futures contract shows congestion with no clear trend signal. This is quasi confirmed by the seasonal which shows a slight bearish signal from June to August. Recommend buying deep out-of-the-money (five steps) $65 strike price puts on feeder cattle with a premium of 50 cents per hundredweight.

Strike price	$65.00/cwt
Premium	$0.50/cwt
Cost of hedge	$0.05/cwt
Estimated option target floor price	$64.47/cwt

Buy in-the-money one-step corn and soybean meal call options because the technical bar chart forecast is slightly bullish for both.

Corn call	Strike price	$2.10/bushel
	Premium	+$0.14/bushel
	Cost of hedge	+$0.04/bushel
Estimated option target ceiling price		$2.28/bushel

Soybean meal	Strike price	$155.00/ton
	Premium	+$8.00/ton
	Cost of hedge	$0.30/ton
Estimated option target ceiling price		$163.80/ton

Figure 8–7   Risk management package #1

meet the feeder's breakeven price of $65 per hundredweight. The broker does not believe that either corn or soybean meal prices will increase, and in fact believes they will decrease; thus, he is proposing no hedge for either commodity. Furthermore, he advises the feeder to sell both corn and soybean meal options and forecasts a net revenue gain for the feeder. This action leaves the corn and soybean meal prices unprotected and thus at risk and further adds more risk to the operation by selling options that have unlimited price risk. As this strategy is unwound, only one of the three risks is hedged and additional risk is added with the selling of the option. Unfortunately, this type of strategy is all too common. Without an RMP to use as a comparison, it is easy to fall prey to a strategy that offers some hedging but also offers a net revenue gain. The RMP provides a structure to compare each action as the strategy is unwound.

Unwinding complex strategies is useful for hedgers as their risks become more embedded and perplexing. The process of building an RMP for each risk situation sharpens the focus for users on why they are considering hedging—to mitigate price risks. The RMP is critical if users are soliciting outside help in creating hedging programs. It is the responsibility of each user to make sure that whatever strategy is offered by an outside consultant ultimately mitigates the user's price risk. The only way to fully do that job is to disassemble the strategy action by action and determine whether or not each action mitigates a risk and how that action compares to the fundamental tools of hedging.

The buyer thinks feeder cattle prices are in a long congestion phase and thus forecasts no major price movement during the feeding period. He recommends buying deep out of the money puts:

Strike price	$64.00/cwt
Premium	$0.10/cwt
Cost of hedge	$0.05/cwt
Estimated option target floor price	$63.85/cwt

He points out that the price floor provides a measure of protection at a very reasonable price ($0.10/cwt × 2,000 = –$200.00).

The broker forecasts that both corn and soybean meal will enter a bearish price period and thus recommends selling naked at-the-money call options.

Sell three corn options @	$0.05/bushel
Sell one soybean meal option @	$5.00/ton

Revenue from corn options = 15,000 × $0.05 = $750

Revenue from soybean options = 100 × $5.00 = $500

Total revenue	$1,250.00
Less feeder cattle put cost	–$200.00
Net revenue gain	$1,050.00

Figure 8–8   Risk management package #2

The Treasury as a Profit Center

A major movement in the 1990s in larger businesses was the movement from the finance department (often called the company's treasury department) being responsible for the day to day financial needs of the business to also being a profit center. The idea that a company could earn additional revenue from just managing their own cash better gained steam in the last half of the twentieth century. In addition to making sure each company division or department had the necessary financial resources to operate effectively, the company's treasury department was expected to also turn a profit. To be sure, this wasn't a new idea. Treasury departments that are well managed have always understood that their financial knowledge could lead to extra revenue and/or reduced costs, yet these actions were generally reported as reducing the cost of borrowed money or increasing the revenue earned from idle cash. The major change in the 1990s was to treat the whole treasury department as a profit center in and of itself with the added responsibility of still being the finance department for the company. The shift to being a profit center can (and did, for many companies) have major consequences.

If the treasury department is doing its job of managing payroll, accounts payable, accounts receivable, fixed capital, and operating capital, the manager of that department and its employees can be evaluated on the performance of how well they get those tasks accomplished. If they happen to also manage their cash flows in such a way as to lower their overall borrowing rate a few basis points in the short run, that would be a nice feather in their caps at performance time. However, the shift of the treasury department to a profit center moves profit to a higher level of importance in performance evaluations. It is not only critical to manage accounts payable properly, but somehow they must also turn a profit. Enter the greed and fear demons and the incentive to take additional risks. Furthermore, as treasury departments take on the added responsibility of earning a profit, they likewise have risks that need to be managed that did not exist before.

Feed Mill Treasury Example

Consider as an example a large agribusiness that has a feed mill division. The feed mill division buys large quantities of grains, oilseeds, and other ingredients and manufactures them into livestock feed which is then sold to various other agribusinesses. The treasury department has the responsibility of managing the accounts payable and receivable. Generally, the treasury department has a weekly deficit of $10 million on average between having to pay for the ingredients and receiving revenues from feed sales. Sometimes the difference is positive and sometimes the deficit is only a minor amount, but the treasury has found that they need to have a maximum of $10 million on hand to handle the deficit. They normally manage this cash flow need with a bank line of credit such that if they don't need the funds, none are expended and thus no borrowing cost, but if they do need the funds, they borrow the necessary amount for a short period and then pay off the line of credit as revenues arrive. This has worked well in the past as funds are borrowed only when needed and generally only for brief periods of time when cash flows from accounts payable for ingredients mismatches with accounts receivable for feed sold.

The company decides to make the treasury department a profit center. Now the manager of the department has to earn income for the division in addition to managing the company's financial needs. The manager knows that the line of credit costs 12 percent when a normal operating loan would be at least 400 basis points lower due to the bank's added responsibility to provide the line and the uncertainty of the amount, thus the bank charges more for the line of credit. The manager knows that on average he borrows 10 million for 6 months total during a calendar year, thus a yearly cost of the line of credit of $600,000 ($10,000,000 × 12%/2). A simple operating loan would be 8 percent, but would be for the full year, thus a total cost of $800,000. The line of credit is still a cheaper alterative. However, the manager finds that a swap dealer will swap the fixed 8 percent debt for a variable rate of LIBOR (London Interbank Offer Rate) plus 1.5 percent subject to adjustment every quarter. The current LIBOR rate is 3 percent. The manager's cost with the swap would be $450,000 ($10,000,000 × 4.5%) versus $600,000 with the line of credit, thus a savings for the treasury department of $150,000.

The astute reader sees that the manager's quick analysis is faulty. The swap deal will provide a rate of 4.5 percent for only one quarter of the year and would then be subject to the change in the LIBOR rate for three more quarters. The LIBOR rate could go down and make the cost of funds even lower than 4.5 percent, or it could increase. Therefore, if the manager accepts the swap deal, the real cost is unknown until the last quarter LIBOR rate is fixed. On the surface, however, the deal appears favorable for at least the first quarter.

The manager needs to complete an RMP. The swap offers the potential for lower operating loan costs but at the cost of an unknown amount of savings and or cost of funds. Without the swap the manager knows his cost of funds to be $600,000 per year. With the swap the first quarter's cost of funds is set at 4.5 percent. The next three quarters have the risk of the LIBOR rate increasing. This increasing rate risk could be hedged by selling short-run interest rate futures contracts or by buying puts.

The manager can now make a more informed decision. He could accept the swap deal and get a lower rate for one quarter of the year and speculate that rates will remain at the same rate or decline. He faces the risk for three quarters of the year that interest rates will increase and cause the swap deal to cost more than the line of credit. Or the manager could take the swap deal and hedge the risk of increasing interest rates and thus manage the risk and potentially yield a cost of funds savings over the line of credit. Since the treasury department is now a profit center, the manager's decision has the added complication of interest rate risk that was not necessarily a major factor to consider in previous assessments.

The manager is now under pressure to produce a profit for the treasury department. One way to do that is to speculate that rates will remain stable or decline. Let's assume the manager takes the swap deal and thus replaces his fixed rate loan for a floating rate loan adjustable every quarter by changes in the LIBOR rate. He saves 150 basis points for the first quarter. Let's further assume that this looks so impressive to the rest of the department's employees that they recommend that he swap not just the 10 million dollars associated with the feed mill operation, but an additional 20 million that they have in floating funds used for payroll creating a substantial first quarter bump in profits. Does this sound like a wise thing to do? The short answer is no. But in fact something similar to this occurred in a famous derivatives example in the 1990s. A large international company entered a swap deal that exchanged their fixed rate loan for a variable rate loan to save less than 50 basis points and swapped double the actual principal amount. Rates increased, and the company found that they were not saving money, but were in fact hemorrhaging cash very rapidly. Short-term greed ended up costing the company millions of dollars. In fact, the company had entered into even more complicated deals with the swap dealer such that neither one fully understood the implications of what they were doing with the end result involving lawsuits, countersuits, and an out-of-court settlement. What the two parties had created was a complex stacked derivative that neither fully understood. A simple RMP with simple risk management tools would have served the company's needs completely and in the process made them a small profit. Greed, ignorance, and complexity for complexity's sake ruined the deal.

A Final Note about Complex Derivatives

Complex derivatives became increasingly popular during the late 1980s and 1990s. As the twenty-first century unfolded, major bankruptcies and less-than-ethical accounting practices combined to cast a pall over any financial transactions that could not be easily understood. All derivative use, either simple or complex, has undergone a more rigid evaluation by management in most companies. As a result, some of the more exotic complex derivatives have vanished from the marketplace. The issue isn't whether or not the derivative is complex or simple. The only issue that matters is the mitigation of a price risk. If it takes a complex derivative to do it, then that is the most appropriate tool to use. If the risk can be handled with a simple derivative, so be it. Occam's Razor is the best concept to adhere to for designing derivatives: "What can be done with fewer, is done in vain with more" (i.e., the idea that the simplest course of action that gets the job done is the best). If a derivative is complex, just to be complex, there is a simpler way to manage the risk and the simpler way is preferred. Albert Einstein once observed, "It is desirable to do things as simple as possible, not simpler."

The major fault of complex derivatives is that they often have both hedging and speculative strategies bound together. The user thinks they are hedged, and in fact may be, but the speculative component of the derivative could have horrendous consequences. The wise user will select the simplest derivative to handle the job and analyze all derivatives via the RMP or some equivalent process to sort out what their costs and benefits really are.

CHAPTER 8—QUESTIONS

1. A cotton merchant sells cotton today for a set price in dollars but will deliver the cotton later and be paid in pesos. What kind of risk(s) does the merchant face? Can those risks be mitigated?

2. Do embedded or cloistered risks require complex derivatives to mitigate risks? Explain.

3. A Midwest farmer produces only corn. His risk is that corn prices will decrease. How might he build a stacked derivative to handle the single risk?

4. Why is it important to have a price forecast before deciding to hedge a risk with either a futures contract or an options contract? Explain.

5. If a complex derivative has both a hedging component and a speculative component, isn't that an efficient management tool?

Managing Other Hedging Risks

The objective of this final chapter is to look at other ideas concerning managing risk, especially the idea of cash streams and supplemental hedging.

OVERVIEW

The previous eight chapters presented a wide range of price risk management tools and strategies, but certainly this array of tools and strategies is not comprehensive. As more, increasingly sophisticated tools are devised by financial managers and derivatives dealers, more, increasingly sophisticated strategies for using the new tools are developed. This chapter highlights a few strategies not yet mentioned that should provide a quick insight into a few of an amazingly large set of standardized tools to accomplish less-than-standard goals and RMPs.

Compensating Balances

One of the reasons for the success of price-based derivatives is that the comparison between the potential cash market loss and the potential gain from mitigating the cash market risk via a price-based derivative on the same commodity is similar—the old idea of comparing apples to apples. A cash market risk of corn price changes is very effectively mitigated with a corn futures contract since they are both price-based risks. Yet the real reason to mitigate a cash risk is not a per unit price per se, but rather the total value that is at risk. A corn producer is concerned with the actual price of corn per bushel but is even more concerned that the total value of corn production be compensated by a risk management tool of like balance—the idea of **compensating balances.**

A corn producer that thinks she will produce 15,000 bushels of corn but only sells one corn futures contract as a hedge is severely under-hedged by 10,000 bushels and will not have compensating balances of risk. Of course, if the action is on purpose so as to only hedge approximately one third of the crop, then that is a viable risk management decision. If the under-hedge is naive, then a better solution is possible. Mismatched balances between the cash position and the proposed mitigating tool are fairly easy to see when they are denominated in the same units such as in bushels. Still the wise risk manager understands that a more complete risk management analysis includes converting compensating balances into **dollar equivalency.**

A corn producer expects to produce 5,000 bushels and suffers a $1,000 loss in the cash market as shown in Figure 9-1. The producer hedged with 20,000 bushels because the hedge was with out-of-the-money options that were cheap and the delta was expected to increase if prices declined. The hedge produced a cash stream gain of $800. The hedge didn't achieve

Cash	Derivative Hedge
Corn crop estimated at 5,000 bushels current rate of $11,000 (corn price, $2.10/bushel)	Buys four put options out of the money at $1.80 strike prices Premiums of $0.02/bu (20,000 bushels)
Sells corn at harvest for $2.00/bu ($10,000 value)	Sells for $0.06/bu
Revenue stream loss $1,000	Revenue stream gain $800

Figure 9.1   Dollar equivalency cash stream hedging

perfect dollar equivalency, but it was 80 percent even though the cash and hedge were mismatched in quantities four to one. All that matters for the hedger is mitigating **cash streams.**

Cross Hedging

Once the idea of dollar equivalency is applied, several other agricultural risks that may not have traditional futures, options, or swaps available become manageable. Hay producers or users have no futures, options, or swap markets available to them to mitigate price risk. What if the price of hay were highly correlated to the price of corn? If hay and corn prices are tightly bound together, the risk of the value of the hay production changing could be mitigated with corn futures or options contracts provided the hedge is constructed to achieve dollar equivalency. Grain sorghum producers and users have used corn futures for years employing this concept. Grain sorghum futures were once traded on the Kansas City Board of Trade, but later were dropped as traders favored the more liquid corn contracts on the Chicago Board of Trade. Similarly, the original mortgage backed security, the GNMA (Government National Mortgage Association), which was the first interest rate futures contract, lost out to the more liquid Treasury Bond contract. Both the feeder and live cattle futures contracts call for steers so heifer producers have no exact derivative to use. Heifer and steer prices are strongly correlated and thus via dollar equivalency risk management is possible for heifers using steer-based derivatives. Figure 9-2 shows a grain sorghum **cross hedge** with corn using the concept of dollar equivalency.

The grain sorghum producer has a revenue loss of $1,600 on 8,000 hundredweight of grain sorghum production. The producer hedged the crop with 10,000 bushels of corn and received a cash stream gain from the hedge of $2,000. What is important to the producer is the dollar stream values of the cash and futures positions, not the actual size of the positions nor for that matter the commodities, only that a loss cash stream was compensated by a cash stream that gained in value.

Hedging Ratios

The action of adjusting the number of derivative contracts to achieve matching cash streams with the cash position is called developing a hedging ratio. Hedging ratios are often developed with past historical information so that past basis movements can be empirically

Cash	Futures
Plants grain	Sells two corn futures (5,000 bushels each)
Sorghum crop estimates 8,000 cwt	for $2.20/bu
Production	
Current price $4/cwt	$22,200 value
Estimated revenue	
$32,000	
Sells at harvest for	Buys @ $2.00
$3.80/cwt and earns $30,400	$20,000 value
Revenue cash stream loss of	Revenue cash stream gain of
$1,600	$2,000

Figure 9.2   Cross hedge of cash streams

adjusted in an attempt to achieve dollar equivalency. Hedging ratios that are developed to compensate for basis changes with futures as the hedging tool are called **risk minimizing ratios.** Properly hedged cash positions with futures contracts have basis risk as the primary risk, and attempts to mitigate that risk via changes in the number of futures contracts used as a hedge result in a hedging ratio other than one for one.

Because futures contracts are standardized, it is difficult to use hedging ratios for small cash positions. If a cash position were only 5,000 bushels of wheat, then a simple futures hedge would be one wheat futures contract—one for one hedging ratio. Only if the hedge period had significant basis variation (risk) would it be possible to have more than one futures contract to minimize the basis risk and thus produce similar cash streams for the cash and futures positions. Figure 9-3 illustrates a simple hedge that has a large enough basis deterioration to warrant having a hedging ratio of two to one. The cash stream from the hedge covered only one half of the cash stream loss from the cash market, thus the necessity of having a larger futures position than the cash position. Obviously for much larger cash positions, the basis variability would have to be less pronounced to justify a hedging ratio greater than one to one.

Supplemental Insurance Hedge
Numerous opportunities exist nationally to buy crop insurance. The general idea is based on some predetermined value that is calculated on historical production records and an average cash market price or some set futures price. **Supplemental insurance hedging** is the idea of increasing the coverage level via derivative. If a corn producer selected a 65 percent coverage level with traditional crop insurance, a supplemental insurance hedge would be an attempt to add a higher level of coverage. Similar consumer products exist where individuals will buy a supplemental health insurance policy to provide protection at a higher level than the basic policy. Supplemental insurance is used for two major reasons: higher level of protection is not available through primary insurance or the added

Cash	Futures	Basis
Buys 5,000 bushels of wheat @ $3.50/bu	Sells one wheat futures @ $3.70/bu	.20
Sells 5,000 bushels of wheat @ $3.30/bu	Buys one wheat futures @ $3.60/bu	.10
Loss of $0.20/bu	Gain of $0.10/bu	
Cash stream loss $1,000 (5,000 × $0.20)	Cash stream gain $500 (5,000 × $0.10)	

Figure 9.3   Hedge with mismatched cash stream flows

cost of higher protection can be achieved with a supplemental policy cheaper through the original policy. Since most forms of crop insurance are revenue based, which includes a production level and a price, it is difficult to hedge the additional supplemental coverage, not because of a price change, but because the production loss is event specific—hail or weather damage. A supplemental insurance hedge would need a derivative that produces a positive cash stream when a weather event produced a negative event. These types of derivatives are not widely used for agriculture and thus supplemental insurance hedging is very difficult.

Just as supplemental insurance hedging is difficult for insurance policies, it is also hard for some governmental programs as well. However, some governmental programs provide, like crop insurance, a certain level of income protection that is not event based, but price based. For additional protection amounts, supplemental hedging is an effective additional risk management tool for these types of programs as illustrated in Figure 9-4. The farmer loses $5,000 in the cash market when he has to put the crop in the CCC loan

Cash	Derivative Hedge
Estimated corn crop value at Today's price $40,000	Buy two $2.50 strike price puts at a premium of $0.10/bu
Crop put into CCC loan program Value $35,000	Sell two $2.50 strike price puts for $0.50/bu
Reserve stream shortfall $5,000	Revenue stream gain $4,000

Figure 9.4   Supplemental insurance hedge for cash strikes

program, but the put option hedge earned a positive cash stream of $4,000 to offset 80 percent of the cash stream loss.

Portfolio Hedging

Portfolio theory for financial products has long been a standard tool to manage certain financial risks. The idea is that several financial products are packaged together such that the risk of a single instrument changing in value is offset with another one that changes in value in an opposite way. Stock portfolios are the most popular because they are a collection of individual stocks such that if one company's stock value decreases, there will be another one that will gain in value. The risk of an individual stock value change is mitigated via the bundling process.

Farmers have long understood this and several manage the risk of one crop changing its value by having another crop that will move in an opposite way. The old agricultural model of *crop and livestock* farming was a simple agricultural portfolio. If crop prices were down, hopefully livestock prices would not be, and vice versa. Portfolio theory is a large and complicated field but the basic ideas are very helpful for simple risk management. Items are included in a portfolio that have no relationship or an opposite relationship to other items in a portfolio. A farmer who grows both corn and grain sorghum does not build a portfolio that will mitigate any significant price risk since corn and grain sorghum prices are highly correlated. A corn farmer that also farms winter wheat might gain some price risk management if corn prices and winter wheat prices don't track together.

A simple corn and soybean farming operation in the Midwest is a portfolio of two crops that are traditionally produced together and don't have a strong positive price correlation with each other. A **portfolio hedge** would be a stacked complex derivative consisting of a mix of risk management tools on corn and soybeans such that the cash stream is protected with a derivative cash stream.

Weather Derivatives

Agriculture is strongly impacted by weather and various tools have emerged in the last decade to help mitigate the risk of weather change. Futures, options, and swaps exist on heating and cooling degree days. Additionally, several private firms have created individual contracts concerning moisture and frost dates that help manage the risk of weather. **Weather derivatives** are most popular with energy companies and utilities, and to a lesser extent with agriculture. Because agriculture has both governmental price support programs and price insurance from private companies that help manage revenue risks to agriculture from weather-based problems, there has been less of a push for weather derivative use in the industry. Granted, while a farmer may not have a strong need to use weather derivatives because he can buy crop insurance, the insurance company does. Insurance companies can absorb the risk or they can reinsure or they can mitigate the risk via active risk management with weather derivatives.

Currently only heating and cooling degree days have futures contracts. Much work needs to be done on correlating heating and cooling degree days with crop and livestock yields for individual areas before these futures will be of any use as a risk management tool. To be sure, heating and cooling degree days are very important for agribusinesses that use large amounts of energy for processing or housing animals. The heating and cooling degree day futures would be helpful in controlling energy costs.

A Final Word on Hedging

All types of price risk management be they futures, options, swaps, or forward cash contracts involve the trade-offs between cash streams and the risk associated with each stream. The analysis of each cash stream can be simple or complex, but must address the fundamental risk associated with each cash stream. A wheat farmer in Kansas has a very risky cash stream based solely on the price of wheat and the variability of wheat production. A food processor that is selling internationally may have not only the risk of wheat prices but foreign exchange and interest rate risks as well and will need more complex derivatives to manage the cash stream of food sales. An orange juice processing plant has the risk of orange juice price changes and also the risk of energy price changes as well. If they get a forward cash contract for energy prices that is tied to orange juice prices (which are available), what was once two different risks (orange juice and energy) has been converted to only one commodity risk (orange juice). If the risk is that orange juice prices will fall on the product side, but increase on the energy side, then they have opposite cash streams that may cancel out. The orange juice processor may not need to use futures, options, or swaps, but may need to make sure the size of the cash streams has dollar equivalency. What is important is that the risk is managed, not the tools used.

CHAPTER 9—QUESTIONS

1. What factors would be important for a corn farmer to consider in estimating the cash stream loss so that a hedge could be constructed?

2. Delta is critical for option hedging and even more important for achieving dollar equivalency. Why?

3. If a farmer can get a governmental price support, isn't that enough cash stream protection? Explain.

4. How is a corn and cattle producer similar to a stock portfolio?

5. If crop and livestock operations were once popular because they helped mitigate price risks, why have most commercial farming and livestock operations become separated during the last 40 years?

Glossary

A

acceptance—absorption of the full production and economic consequence of an uncertain outcome.

arbitrage—the process of simultaneously selling and buying in two or more markets to take advantage of a perceived difference.

at-the-money—a put or call option whose strike price is the same as the price of the underlying contract.

B

backwardation—cash price higher than futures price.

bar chart—a tool of technical analysis that measures the high for the day and the low for the day such that the image appears as a vertical line with a bar denoting where the market closes or settles.

basis—the difference between a cash price and a futures price.

basis trade—futures contract trading process that involves the relative difference between the futures market and the market for the underlying cash commodity.

bear—a trader who believes the market price will fall.

bid-ask spread—the sum received by a swap dealer in payment for his/her services in arranging the swap. It is the difference between the amount received by the swap dealer from one counterparty and the amount paid to the other counterparty.

biological risk—deviations from expected production due to biological causes such as weeds, pests, diseases, or biological reactions in a production process.

blow up—a catastrophic loss in which a trader loses more than he was able to withstand.

brokers—individuals who act on behalf of traders to buy or sell a futures or options contract.

bull—a trader who believes the market price will rise.

butterfly—a trade that simultaneously involves the purchase of two different futures delivery months and the selling of two of the same delivery months between the two purchases, as: buy one May corn futures; sell two July corn futures; buy one September corn futures. The butterfly trade is used to take advantage of nearby prices strengthening relative to faraway prices. *See* Spreading, Arbitrage.

buy hedge—a bull hedge or a long hedge. A hedge entered into to offset the impact of a rising price.

C

call—an option that the buyer has the right but not the obligation to purchase the underlying contract or commodity.

call contract—a basis contract that requires the contract holder to notify the contract provider and supply information about when the contract will be exercised or completed.

candlestick chart—a variation of a bar chart that measures the opening and closing price as well as the high and low for a trading period.

cash settlement—the process of discharging or offsetting a futures contract that has expired. The futures obligation is offset by calculating the difference between the final futures price and a final cash price (usually an average or index value).

cash streams—projected cash flow pattern due to a particular position in any market.

caveat emptor—"let the buyer beware."

caveat venditor—"let the seller beware."

chart analysis—*See* Technical analysis.

clearinghouse—a third party between traders that assures the performance of each trader.

cloistered derivative—a complex derivative comprised of several risk management tools to mitigate a bundle of risks of embedded or cloistered risks.

cloistered risk—a risk that is embedded in another risk. The risk of the value of a finished product changing has the embedded risks of the price of the inputs changing.

commodity futures trading commission (CFTC)—the federal regulatory agency for all futures and options contracts traded on organized exchanges. The CFTC was created in 1974 to replace the Commodity Exchange Authority. The CFTC is composed of five members appointed by the President and confirmed by the Senate.

commodity swap—swaps that involve the exchange of cash flows due to positions taken in an energy or metals market.

compensating balances—the process of creating dollar equivalency between the cash position and the derivatives used to hedge the position such that the cash position is fully protected.

complex derivative—a position taken in a futures or options market to hedge a position created as a result of a swap.

contango—futures price higher than cash price.

cost of carry—the costs involved in storing from one time period to another. Usually is composed of actual storage costs, insurance, and the time value of money.

counterbalance—characteristic of two markets that tend to move together.

counterparty—one who enters into a swap agreement.

covered option writing—selling call options with the ownership of a long futures position and put options with the ownership of a short futures position. Covered call writers are bullish and covered put writers are bearish.

cross hedging—the process of using a derivative to hedge a cash position such that neither the cash position nor the derivative match exactly such as using corn futures to hedge a cash grain sorghum position.

crush hedges—hedges involving a commodity which will be processed into one or more different commodities (such as soybeans processed into soybean meal and soybean oil). A positive crush is placed when the values of the products exceed the cost of the input plus the cost of processing.

D

day trader—a person who holds an open position for no longer than a given trading day.

daily trading limit—the price limit, both high and low boundaries, that the futures price must trade within each day.

default risk—the risk that a contract will not be fulfilled as specified.

delivery—the process of offsetting an open futures position by going through the process of physical delivery or cash settlement.

delivery point—the designated place the commodity must be moved to in order to satisfy the terms of delivery.

demutualization—the process of converting a mutual business to another business form, usually to a conventional stock company.

derivative—a financial instrument whose value is obtained from another source.

dollar equivalency—equalizing balances (compensating balances) between two positions of a hedge in dollar terms (as opposed to tons or bushels or some other physical units).

double derivative—a contract that draws its value two steps removed from the original source such as an option contract that derives its value from a futures contract which in turn derives its value from the underlying cash commodity.

E

econometrics—the use of statistics and mathematics to evaluate economic theories.

effective date—the beginning date of a swap; the date the swap takes effect.

efficient market hypothesis—a theory that markets contain all public and knowable information and therefore cannot be forecasted with any accuracy.

exchanges—organizations that provide central trading facilities and/or processes.

exercise—the conversion of an option into a demand for performance on the underlying contract.

electronic trading—centralized trading using computer systems rather than physical facilities.

F

financial engineering—managing risk through the use of sophisticated risk management tools and strategies with advanced computer analyses using substantial amounts of financial information.

financial management—same as financial engineering.

fixing—reference price used to set the terms of trade for a swap contract.

floor broker—a holder of a seat on an exchange who trades in the pit for customers that issue orders.

floor trader—a holder of a seat on an exchange who trades in the pit for his own account.

forward sell—the act of selling an item for future delivery.

fundamental analysis—the process of understanding the underlying cause and effect factors that cause prices to move or to reach equilibrium.

futures commission merchant (FCM)—a person who buys/sells futures contracts for a client for a fee.

futures contract—a legal obligation to deliver (a sell) or accept delivery (a buy) of a specific commodity with contract terms standardized.

H

hedge—the process of shifting price risks in the cash market to the futures market by simultaneously holding opposite positions in the cash and futures markets.

I

inefficient market hypothesis—a theory that says not all market information is reflected in the market price, thus there are opportunities to profit from the lack of information, especially by forecasting future price directions.

interest rate swap—a swap that involves loans or bonds; usually an interest rate swap involves the exchange of the cash flow from a fixed rate loan or bond for the cash flow from a variable rate loan or bond.

in-the-money—a put (call) option that has a strike price that is higher (lower) than the price of the underlying contract.

intrinsic value—the numeric difference between an in-the-money strike price and the price of the underlying contract.

inverted market—distant futures delivery month prices are lower than nearby month's price.

L

last trading day—the last day that a futures or options contract can be traded before the contract month expires.

liquidity—a characteristic of markets that indicates sufficient trading activity to allow sellers to sell and buyers to buy quickly.

London Interbank Offered Rate (LIBOR)—rate international banks charge each other for loans and thus a popular rate to use to formula price variable rate loans.

long—an initial buy position of a futures or options contract or the physical ownership of the cash commodity.

long hedge—an initial purchase of a futures contract or an initial purchase of a call option contract used to protect against an initial forward sell in the cash market.

M

maintenance margin—the predetermined amount of the initial margin that triggers a margin call signifying that the position has lost enough money to require more cash to hold the position.

margin—the initial amount of good faith cash that must be posted with a broker to enter into a futures or options position.

margin call—the amount of money that must be deposited with the broker to maintain a losing futures or options position.

marketing risk—deviations from expected sales due to changes in market conditions and/or availability.

maturity—for a swap, the maturity is the time period between the effective date and the termination date; thus, the effective life of the swap.

monopoly—a market structure characterized by one seller.

monopsony—a market structure characterized by one buyer.

moving averages—a charting tool that calculates averages for different time periods such as three days and ten days to smooth out daily variations.

mutual—a business structure in which the users are also the owners.

N

naked option writing—selling an option without owning the underlying futures contract. Naked put writers are bullish and naked call writers are bearish.

national futures association—formed by Congress in 1982 to regulate the futures and options markets via self-regulation with dues from members.

net hedge price—the net price a hedger receives considering both cash and futures transactions.

neutralizing—avoidance or removal of the economic consequence of an uncertain outcome.

notional—the principal value that determines the cash flow value for swap contracts.

O

offset—the opposite action taken to get out of an initial futures or options position.

oligopoly—a market structure characterized by a few sellers.

oligopsony—a market structure characterized by a few buyers.

one-step derivative—a contract whose value is determined from another contract that is once removed from the original source such as a futures contract that derives its value from the underlying cash commodity.

open interest—the number of open (not yet offset) contracts.

open outcry—the process of obtaining an established price for a futures or options contract in the trading pit. It must be both by voice and hand signals.

option—a contract that gives the buyer the right but not the obligation to obtain an item/service. The seller of the contract has an obligation to perform, should the buyer exercise the right.

oscillators—a family of trading tools that use a simple arithmetic expression to measure the rate of change of prices.

out-of-the-money—a put (call) option that has a strike price that is less (more) than the price of the underlying contract.

over-hedged—the amount of product hedged is greater than the expected amount to be sold or bought.

P

paper gain—the amount that the current futures price is different from the initial price such that a gain would occur if offset.

paper loss—the amount that the current futures price is different than the initial price such that a loss would occur if offset.

passing—shifting risk to another person.

pay-receive spread—same as the bid-ask spread.

perfectly competitive—a market structure characterized by many buyers and many sellers. No one buyer or seller is sufficiently large to affect price.

pit—the actual place where open outcry occurs and the futures and options contracts trade at the exchange.

plain vanilla swap—a basic swap; a simple swap of two cash flow streams between two counterparties.

point and figure chart—a technical trading tool that measures major price directions without regard for time.

policy risk—deviations from expected financial results due to changes in public policy, including agricultural subsidies, tax law, trade organization decisions, or court decisions.

position limit—the maximum number of contracts that a trader can own and/or control of each commodity.

portfolio hedging—hedging a group (or portfolio) of commodities or stocks with one instrument or strategy.

premium—the price that an option contract trades for a given strike price.

price elasticity of demand—a measure of how responsive the quantity demanded is due to a change in the price of the product.

price elasticity of supply—a measure of how responsive the quantity supplied is due to a change in the price of the product.

price insurance—using options contracts to insure or protect against adverse price movements while retaining profits from favorable price movements.

probability—a quantitative measure of uncertainty.

production risk—deviations from expected production levels; can be caused by a myriad of factors, including weather, insects, and diseases.

put—an option contract that the buyer has the right but not the obligation to have a sell position on the underlying contract.

R

random variable—a numeric value that occurs by chance.

random walk hypothesis—the theory that underlies the Efficient Market Hypothesis that states all information moves in and out of the market as a random variable.

reference price—price used to determine the value of the cash flows exchanged with a swap contract.

regulation—oversight of market activities and functions conducted by legislatively-authorized public institutions such as the CFTC.

reverse crush hedges—a hedge placed when the value of the products are less than the cost of processing plus the cost of the input.

reverse spread—a simultaneous position that involves having a sell position on the nearby futures contract and a buy position in a more distant futures contract.

risk—an unknown outcome that impacts a business or individual.

risk averse—a characteristic of a person who attempts to avoid, displace, shift, or pass risk to another.

risk lover—a characteristic of a person who attempts to profit by accepting risk from another or from the market.

risk minimizing ratios—hedging ratios developed to compensate for basis changes when using futures contracts.

Risk and Mitigation Profile (RMP)—an assessment of the risks being faced by a person or business including a determination whether those can or should be neutralized or passed.

risk neutral—a characteristic of a person who is indifferent to the consequences of risk; neither avoids nor seeks risk.

rolling hedge—the process of offsetting the futures position and replacing with a more distant futures position because the cash position has not been altered.

roundturn—*see* Offset.

S

scalper—a trader in the pit that holds a position only briefly and trades on small price moves.

seat—a membership at an exchange that gives the owner or leasor the right to trade at that exchange.

second best—the next best outcome. Options are often called second best—if prices move against the position taken, the trader would have been better off hedged with a futures contract and if prices move in favor of the position taken, the trader would have been better off unhedged.

selective hedging—deciding whether to place a basis hedge.

sell hedge—a bear hedge or a short hedge. A hedge entered into to offset the impact of a falling price.

service payment—the amounts of money exchanged between the two counterparties involved in a swap.

settle price—the final price used each day to value futures and options contracts. It involves the actual last traded price and a weighted average of the last trades during the last few minutes of the trading day.

short—an initial sell position with a futures or options contract or the act of forward selling a cash position.

short hedge—An initial sell in the futures market to offset a long cash position.

single risk—risks that are easy to separate and keep separate. If the risks cannot be kept separate, they become embedded or cloistered risks.

speculator—someone who trades futures or options contracts with the intention of making a profit and does not own or control the actual cash commodity.

spot market—the actual cash market at the current moment.

spread—the simultaneous position involving an initial buy of the nearby futures contract and an initial sell of a more distant month.

stacked derivative—a complex derivative comprised of several risk management tools to mitigate a bundle of single risks.

standard deviation—the square root of the variance and the generally accepted measure of risk.

standardized contract—a contract with prespecified uniform terms concerning size, grade, and delivery among other terms; usually these terms are established by an independent third party exchange.

stop—the placing of an order to offset an initial futures position if a certain price level is reached.

strike price—the price that an option contract will be converted into the underlying position.

strong contract—a contract that can be retraded.

supplemental insurance hedging—the process of using derivatives to provide an added level of price protection when other forms of cash protection (such as crop insurance) do not provide the desired level of coverage.

swap—a contract involving two or more counterparties that agree to exchange cash flows.

synthetic futures hedges—the process of buying a put (call) and selling a call (put) naked to mimic a short (long) futures position.

T

tandem—two actions taken together to remove a price risk.

target price—the estimate of a net hedge price.

technical analysis—the belief that futures price direction can be determined by past price movements.

termination—the ending date of a swap; the date the swap terminates.

tick—the least value change that can occur in the price movement of a futures or options contract.

tier or ladder maturity swap—swaps that must be sequenced to discrete events such as an annual harvest.

time value—the amount of the option premium that reflects the trader's expectations of future value. The net difference between the intrinsic value and the option premium for in-the-money options. Time value and the option premium are the same for at-the-money options and out-of-the money options.

time value of money—the opportunity cost of money.

transference—a shift of the economic consequences of an unknown outcome from one market to another.

two-step derivative—a contract whose value is determined from another contract that is twice removed from the original source, such as an options contract that derives its value from a futures contract (a one-step derivative) that derives its value from the underlying commodity.

U

uncertainty—an unknown outcome.

under-hedged—the total amount of product hedged is less than the expected amount to be sold or bought.

underlying contract—with options on futures contracts, it refers to the futures contract. With options on the actual commodity, it refers to the actual commodity.

unwinding—the process of dissembling a risk management package (or a risk mitigation profile).

V

variance—a parameter that measure the dispersion of a distribution.

volume—the number of total contracts traded.

W

weak contract—a contract that cannot be retraded.

weather derivative—futures contracts and strategies based on temperature (such as frost-free or heating degree days) or precipitation.

weather risk—deviations from expected production caused by unanticipated weather events.

writing—selling an option contract.

Index